AF335238

Geophysical References, Volume 4
Marion R. Bone, Series Editor

Seismic Wavefield Sampling

A Wavenumber Approach to Acquisition Fundamentals

Gijs J. O. Vermeer

Edited by Michael R. Cooper

Society of Exploration Geophysicists

Post Office Box 702740 / Tulsa, Oklahoma 74170-2740

Vermeer, Gijs J.O.
    Seismic wavefield sampling : a wave number approach to acquisition
fundamentals / Gijs J. O. Vermeer ; edited by Michael R. Cooper
        p.   cm. — (Geophysical references ; v. 4)

    Includes bibliographical references and index.
    ISBN 0-931830-47-8 (series) : $70.00. — ISBN
    1-56080-010-0 (volume)

    1. Seismic waves. 2. Seismic reflection method. 3. Seismic
prospecting. I. Cooper, Michael R., 1947–   .
III. Title. IV. Series.

QE538.5 V47   1990   90-10215
551.2'2'0287-dc20   CIP

ISBN 0-931830-47-8 (Series)
ISBN 1-56080-010-0 (Volume)

Published 1990 by the Society of Exploration Geophysicists.
Second printing 1992

Printed in the United States of America

To my parents

Contents

Nomenclature

$A(t)$ = antialias filter

$B[x_m]$ = line length box function

$B[x_o]$ = spread length box function

C = a constant

CHP = common horizon panel (horizon time slice)

CMP = common midpoint panel = (t,x_o) panel

COP = common offset panel = (t,x_m) panel

CRP = common receiver panel = (t,x_s) panel

CSP = common shot panel = (t,x_r) panel

CTP = common time panel = (x_s,x_r) or (x_m,x_o) panel

d = shot or receiver sampling interval (station spacing)

d_r = depth of receiver

d_s = depth of source

dx_r = receiver sampling interval

dx_s = shot sampling interval

f = frequency

f_{max} = maximum frequency

f_N = Nyquist frequency (= $1/2\Delta t$, usually = f_{max})

$g(t,x_i)$ = two-dimensional filter

$G(t,x_s,x_r)$ = ghost filter

$h(x_i)$ = spatial filter, $i = s, r, m, o$

$j = \sqrt{-1}$

k_i = wavenumber corresponding to x_i, $i = s, r, m, o$

k_N = Nyquist wavenumber (= $1/2d$)

$|k_i|_{max}$ = maximum wavenumber k_i corresponding to f_{max} and V_{min}

m = subscript for midpoint

$n(t,x_s,x_r)$ − additive ambient noise

NMO = normal moveout

o = subscript for offset

p = apparent slowness in (τ,p) transform

p_i = $\partial t/\partial x_i$ apparent slowness of event in (t,x_i),
  $i = s, r, m, o$

p_r = receiver pattern response

p_s = shot pattern response

r = (1) subscript for receiver
  (2) reflection coefficient

$R(t,x_r)$ = receiver response (geophone or hydrophone)

s = subscript for shot (or source)

$S(k_o)$ = stack response

S = stretch factor

$S(t,x_s)$ = source wavelet

$S_{tot}(k_m,k_o)$ = total response of field patterns and stack

S_Δ = sampling operator

t = (1) recording time, traveltime
  (2) arrival time of event
  (3) time in time filter

$t(x_s,x_r)$ = time function (event or surface) in (x_s,x_r)
t' = physical time
$u(t,x_s,x_r)$ = the recorded wavefield
$U(t,x_s,x_r)$ = a processed version of $u(t,x_s,x_r)$
V = seismic velocity
V_{app} = apparent velocity
V_i = apparent velocity in (t,x_i), $i = s, r, m, o$
V_{min} = minimum phase velocity of events
V_{NMO} = normal moveout velocity or stacking velocity
V_R = Rayleigh wave velocity
V_S = shear wave velocity
V_w = seismic velocity in water
$W(t,x_s,x_r)$ = continuous wavefield
W_i = Fourier transform of $W(t,x_s,x_r)$ over i
W_{ij} = Fourier transform of $W(t,x_s,x_r)$ over i and j
W_{ijk} = Fourier transform of $W(t,x_s,x_r)$ over i, j, and k

$\left.\begin{array}{l} \\ \\ \end{array}\right\}$ $i, j, k = x_s,x_r,x_m,x_o$ or t

(x,y,z) = spatial coordinates
(x_r,y_r,z_r) = receiver coordinates
(x_s,y_s,z_s) = shot coordinates
x_m = midpoint coordinate
x_o = offset = (signed) distance shot-receiver
Δ = distance between pattern elements
Δt = (1) time sampling interval
         (2) small time interval
Δx_i = basic sampling interval in x_i, $i = s, r, m, o$
ω = radial frequency
μ = shear modulus
ρ = density
τ = intercept time
Φ_i = (τ,p) transform of $W(t,x_s,x_r)$ over i, $i = x_s, x_r, x_m, x_o$
Ψ = double (τ,p) transform of $W(t,x_s,x_r)$

Notes (1) Some subscripts may be used in combination, e.g., k_{s_N} is Nyquist wavenumber for shot coordinate x_s,

(2) A comma between subscripts indicates 'or', e.g., $k_{s,r}$ means k_s or k_r.

Acknowledgments

I am most grateful to my wife, Tini, without whose continuous encouragement and support this book would not exist. Numerous people with whom I have discussed the subject matter of this book include particularly M. van Rees and J. G. A. Hillaert who helped formulate some early ideas; R. S. Simpson who shared his ideas on weighted stacking; and W. B. M. Roosdorp, P. R. Zaanen, R. Hoogenboom, A. R. Verdel, R. W. Calvert, A. M. G. Delsing, and R. Baseliers who contributed many varied suggestions. I also express gratitude to the many people that helped with figures and text.

The manuscript submitted to the SEG, was reviewed by Mark J. Graebner and Wim E. Peet. Their many suggestions helped reformulate a great number of paragraphs.

Michael R. Cooper, the book editor, showed how often I jumped from A to D, leaving the derivation of steps B and C to the reader. His comments helped improve clarity considerably.

Final editing and preparation of the book for printing was carried out by Lynn Griffin of SEG's Business Office. Not many paragraphs in this book escaped her skillful corrections.

Finally, I thank Shell Internationale Petroleum Maatschappij for granting me permission to publish this book, and allowing the use of seismic data from various sources.

Preface

When seismic data acquisition changed from single coverage to multiple (mostly 6-fold) coverage in the 1950s, a third dimension was literally added to the data recorded for the 2-D seismic line. However, with 6-fold data this third dimension is still difficult to recognize. The fact that the changeover from 6- to 60- or 120-fold took some 25 years may explain why the third dimension took a long time to be perceived by most practicing geophysicists.

Various publications introduce and use the relation between the two three-dimensional coordinate systems (shot/receiver and midpoint/offset) that describe the prestack data set (e.g., Claerbout, 1985, Hampson, 1987). However, these notions have never resulted in publication of a textbook covering the fundamentals of seismic data acquisition and processing. This book provides theories and some practical recommendations needed for an optimum seismic data acquisition technique, with its main thrust concentrated on spatial sampling.

This book is not a discourse on the theory of seismic wave propagation but deals with the properties of the seismic wavefield only insofar as they are observed by the receivers in seismic data acquisition. Neither are the more practical aspects of seismic data acquisition, such as choice of instrumentation and choice of sources and receivers, covered.

The earliest version of this book was written in 1979–1981 during my employment at Nederlandse Aardolie Maatschappij (NAM) in Assen, The Netherlands. The writing started with a derivation of the equivalence theorem, that states under which conditions (f,k) filtering before or after stack would produce near identical results. The derivation of those conditions gave me better insight in the relationships between the four spatial domains: common shot, common receiver, common midpoint, and common offset.

Perhaps the greatest stimulus to develop the theories described came from the work of Martin van Rees, who was seismic acquisition geophysicist at NAM in the years around 1980. In 1980 he concluded in an internal Shell report:

> "Our studies and practical experience show that an equal sampling in shot and receiver domain is a (near) optimum solution. In practice this means a shot spacing equal to the receiver spacing for off-end shooting and twice the receiver spacing for a split-spread geometry."

Van Rees' conclusion comes very close to the *symmetric sampling* technique demonstrated here as the theoretically optimal technique.

Van Rees' report demonstrated the independence of shot and receiver domains using displays of common receiver panels after common shot filtering, and of common shot panels after common receiver filtering. Similar displays are shown in Figures 1.1 through 1.4 of this book.

With the results obtained by Van Rees as a starting point, I developed the basic theories behind the acquisition and processing of the 2-D single seismic line. I found out about the relevance of reciprocity, the relation between minimum apparent velocity and sampling, and the effect of spatial filtering in various domains. The results of this work were published in 1982 in an internal Shell report.

In 1986 *The Leading Edge* published Anstey's important article, "Whatever happened to groundroll?", which described the stack-array approach. Anstey presumed that his approach was not new, and that some companies in the industry might already practice similar ideas. At least for Shell he was right, as is clear from the above quote from Van Rees' work, from published work of Ongkiehong and Askin (1988), and from this publication. Anstey deserves much credit for being first to discuss some new ideas on seismic data acquisition.

Also, as a result of the discussion that followed Anstey's article, my understanding of the subject matter matured considerably and many refinements and discussions of articles published by various authors in the 1980s that are included, were absent in the Shell report.

Many results described were published earlier in different contexts by other authors. This book brings these results together, synthesized in a framework of simple mathematical considerations, and amalgamated with quite a few new ideas. The book should interest both experienced exploration geophysicists and novices.

Gijs J. O. Vermeer

References

Anstey, N., 1986, Whatever happened to groundroll? Leading Edge, **5,** no. 3, 40–45.

Claerbout, J. F., 1985, Imaging the Earth's Interior: Blackwell Scientific Publ.

Hampson, G., 1987, The relationship of pre-stack apparent velocity filtering to the symmetry of the CMP stack response: First Break, **5,** 359–377.

Ongkiehong, L., and Askin, H., 1988, Towards the universal seismic acquisition technique: First Break, **6,** 46–63.

Chapter 1
INTRODUCTION

The use of reflection seismic for the exploration of hydrocarbons entered a new epoch with the introduction of the horizontal stacking method in the early 1950s (Mayne, 1962, Sheriff and Geldart, 1982). In his pioneering article Mayne explained the benefit of using shorter receiver patterns for the preservation of reflected signals while delegating some of the groundroll suppression to the stacking process.

For a long time the 6-fold stack was the ultimate of the seismic method, due to limitations in acquisition and processing hardware. Not until the late 1970s did reductions in station spacing, combined with a large increase in the number of recorded channels, allow higher stacking multiplicities leading to further drastic improvements in the quality of the final seismic reflection sections.

The introduction of multiple-coverage recording called for the use of multichannel processing. Stacking, the multichannel process introduced first, was soon followed by poststack multichannel processes such as velocity filtering (Fail and Grau, 1963; Embree et al., 1963) and migration.

Prestack multichannel processes became more feasible and even mandatory when the increase in number of recorded channels led to larger multiplicities and the reduced pattern lengths resulted in lower signal-to-noise (S/N) ratios. Some multichannel processes developed in the 1970s and 1980s are: multiple elimination, surface consistent statics correction, surface consistent deconvolution, slant stacking, dip moveout (DMO) correction, and prestack migration. Most, if not all, of these techniques perform best for high multiplicities and regular spatial sampling.

Though a variety of sophisticated multichannel processing techniques were developed by the best scientists in the industry, not much attention has been paid to the basic theories required for an optimal definition of acquisition parameters. Progress in the field of data acquisition has been driven by technology, experimentation, and intuition, rather than by sound theories.

A breakthrough in the formulation of seismic data acquisition techniques came with Nigel Anstey's papers on the stack-array approach (1986a and b). Anstey's recommendations, which were based on experience and intuition, need to be modified and extended on the basis of a more theoretical approach to the three-dimensional aspects of multiple-coverage data. This theoretical background is provided here and the consequences for seismic data acquisition and seismic data processing are elaborated upon.

The reader's basic knowledge of the seismic method and of signal processing is assumed. Sheriff and Geldart (1982) is recommended as a more general introduction to seismic exploration. Some aspects of offset as the third dimension of multiple-coverage two-dimensional (2-D) seismic data are also covered in Claerbout (1985).

A better understanding of the three-dimensional nature of the wavefield of the seismic line will help answer such questions as:

—Why do steep events seem not to be affected in the common shot panel after (f,k) filtering in the common receiver panel and in the common receiver panel after (f,k) filtering in the common shot panel (Figures 1.1–1.3), and why are they suppressed best by application of both shot and receiver filters (Figure 1.4)?

—What are the limits of seismic data quality improvement after so much has already been achieved?

This text starts with spatial considerations in Chapter 2 that describe some basic results on signal processing in the spatial coordinate for the two-dimensional stacked seismic line.

Chapter 3 introduces the three-dimensional continuous wavefield (assumed to be scalar for simplicity) of the 2-D single seismic line which is the hypothetical wavefield that would be generated by the ensemble of all possible shots, each recorded by an infinite number of infinitely closely spaced receivers. Although hypothetical, the wavefield properties need to be known to enable a proper selection of data acquisition

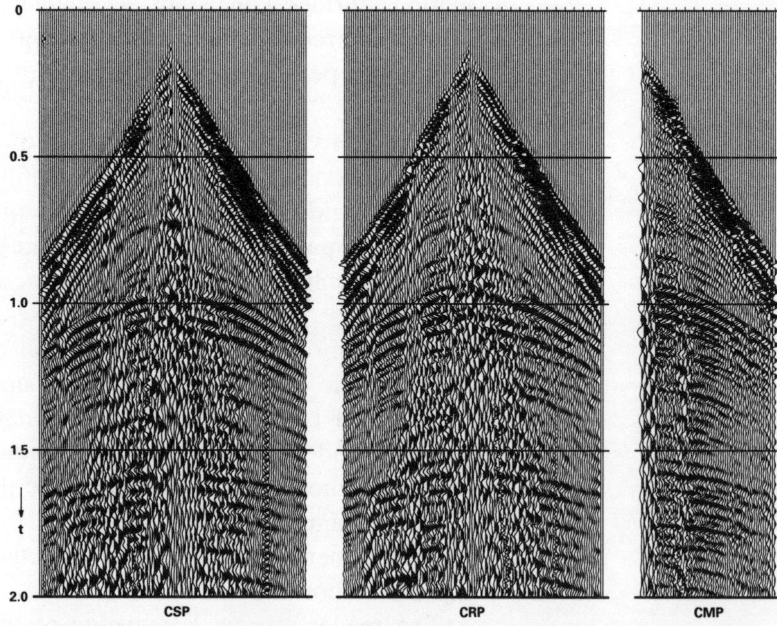

Fig. 1.1. Input data used for experiments with shot and receiver (f,k) filtering. See Nomenclature for abbreviations.

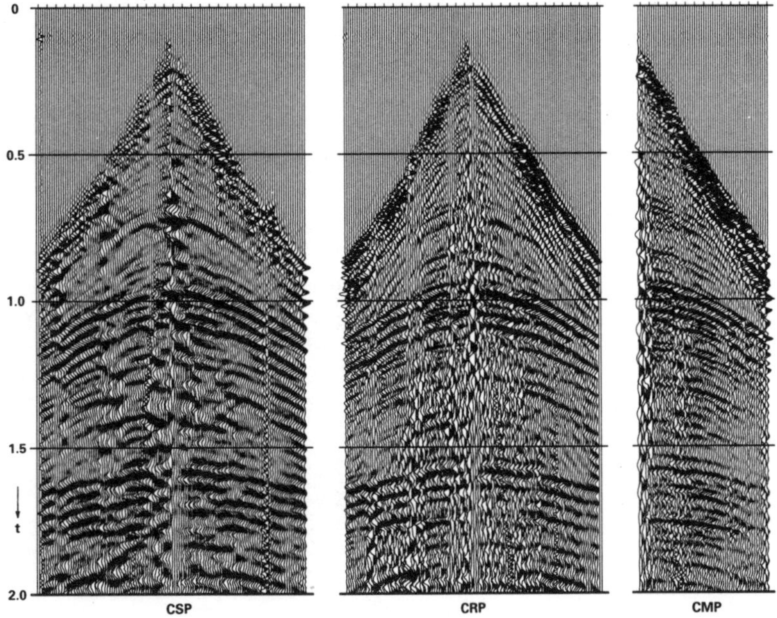

Fig. 1.2. Data of Fig. 1.1 after common shot (f,k) filtering. Though all traces have been modified, the groundroll is still conspicuous in CRP and CMP. (For discussion of Figures 1.1–1.4 see Section 4.10.2.)

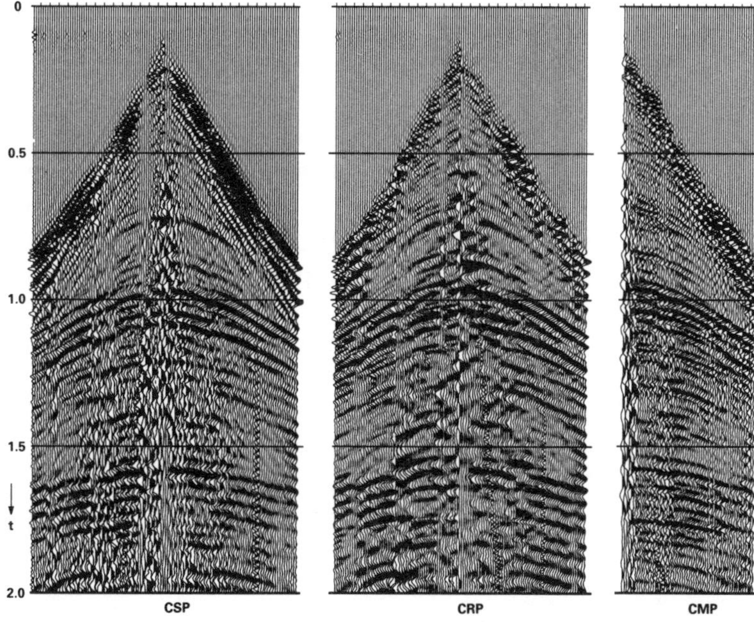

Fig. 1.3. Data of Fig. 1.1 after common receiver (f,k) filtering. Though all traces have been modified, the groundroll is still conspicuous in CSP and CMP.

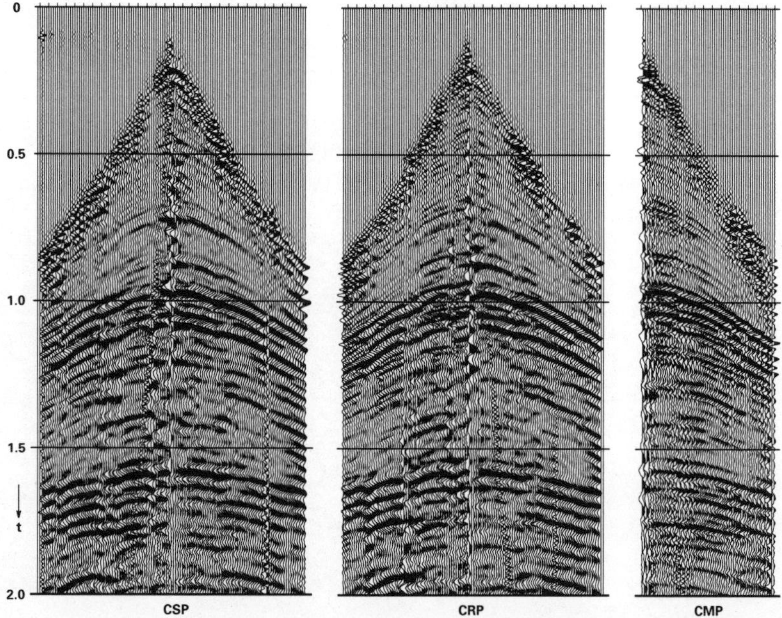

Fig. 1.4. Data of Fig. 1.1 after cascaded application of common shot and common receiver (f,k) filters. Note improved groundroll suppression in the CMP.

parameters. Next, the relationships between the spatial coordinates, shot/ receiver coordinates, and midpoint/offset coordinates are derived. Apparent velocity in the various domains is discussed, and the principle of reciprocity is related to the continuous wavefield. The chapter ends with a number of examples of events in the three-dimensional prestack wavefield.

Chapter 4 discusses the recorded wavefield. A derivation of sampling criteria for the various spatial coordinates is given, and much attention is paid to shot and receiver patterns, their interrelationship, and their influence on data quality. That a symmetric sampling technique is required for optimal results is demonstrated.

Chapter 5 deals with the processing of the recorded wavefield but only insofar as it bears a direct relation to the theories developed in Chapters 3 and 4. Much attention is given to stacking and its relation to the field technique.

Chapter 6 summarizes some important points discussed in earlier chapters.

Chapter 2
SPATIAL CONSIDERATIONS

From the early days of seismic data acquisition, seismic wavefields (continuous time functions) were sampled at discrete x positions along a seismic line. Since the introduction of digital recording the time function is also sampled at discrete intervals. Still we are inclined to think of a seismic section as a collection of continuous time functions. This thinking results from the conventional display of a seismic section as a series of continuous traces at regular spatial intervals.

The data samples of a seismic section, for instance a stacked section, form the digital representation of a two-dimensional function that is basically continuous in t and in x (Figure 2.1). The samples, or data points corresponding to a constant x, are usually displayed as a continuous function of traveltime t (a trace), but we could just as well display data points corresponding to a constant t as a continuous x trace as is illustrated in Figures 2.2a and 2.2b. Both displays represent exactly the same data. (Usually the time scale is more compressed, Figure 2.2c.)

We are used to processes applied to the individual traces: bandpass filtering and deconvolution. Similar processes can be applied to constant time traces. The spatial spectrum of a seismic section can be analyzed using an exclusive wavenumber filter test in analogy to an exclusive frequency filter test (Figure 2.3). The zero-phase wavenumber filters may consist of exactly the same filter points as the corresponding frequency filters. In fact, the example in Figure 2.3 was created by first multiplexing the data to constant-time traces and then applying a frequency-filtering program (followed by demultiplexing). Note the effect of cutting out the low-wavenumber energy in Figure 2.3c: the horizontal and subhorizontal events disappear. This effect is quite different from the effect the same filter would have had on constant x traces. In those traces the dc component is already absent and the filter would remove only high frequencies. Figures 2.3e through 2.3h reveal some crisscrossing events that correspond to irregularities in the unfiltered section in Figure 2.3a. Whether these events are an artifact of migration or whether they have some other origin might be established by running the same filter test on the stacked section.

Resampling to a larger sampling interval is a frequently applied operation on constant x traces. We know that such a procedure should be preceded by the application of an antialias filter. The same procedure should be followed when reducing the number of constant x traces (sometimes called binating or decimating). However, common practice still is to reduce the number of (constant x) traces by a simple mix, or even by selection of every nth trace, procedures nobody would accept for resampling in t. Figure 2.4a shows a data set reduced from a 6 times larger set without a preceding antialias filter. Figure 2.4b shows the result when the same operation was performed after application of a

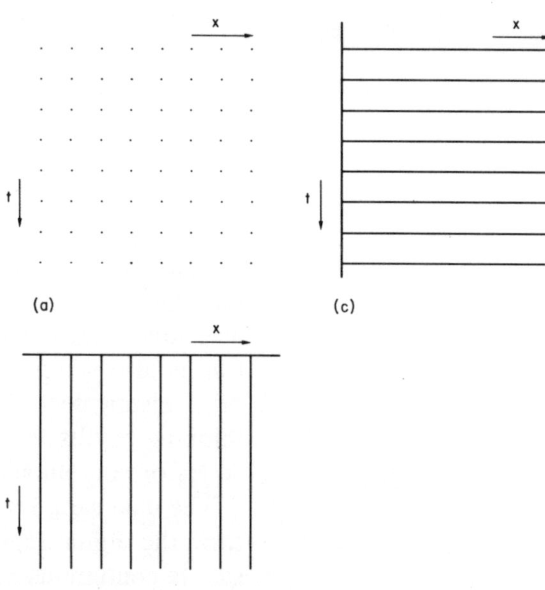

Fig. 2.1. A seismic section is a display of a two-dimensional collection of datapoints (samples). (a) Matrix of data points, (b) Usual representation is a continuous display in t, at discrete x's, (c) Alternative representation is a continuous display in x, at discrete t's.

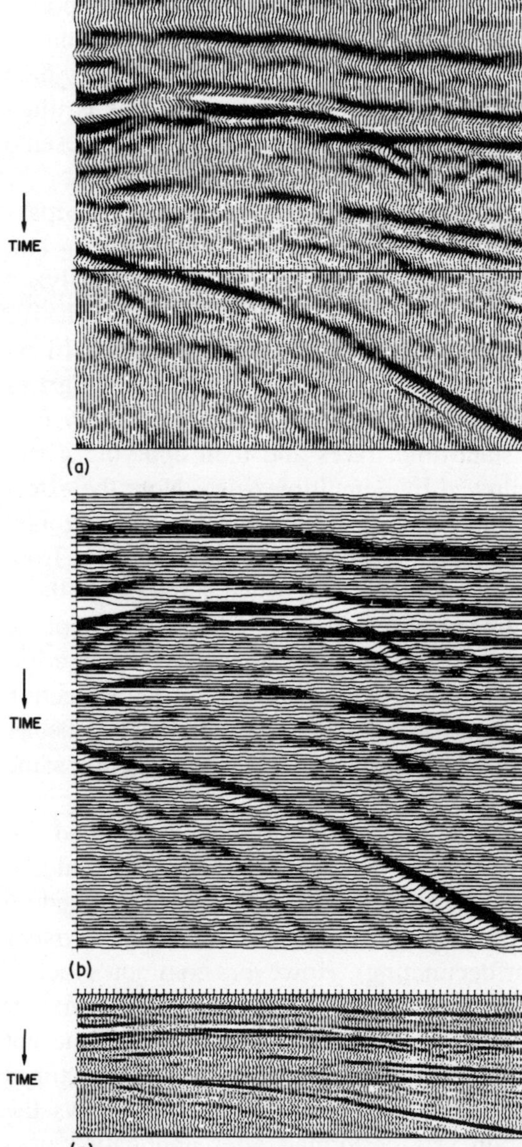

Fig. 2.2. The same data displayed in three different ways. (a) Square grid of vertical traces, (b) Square grid of horizontal traces, (c) Conventional display with compressed time-scale.

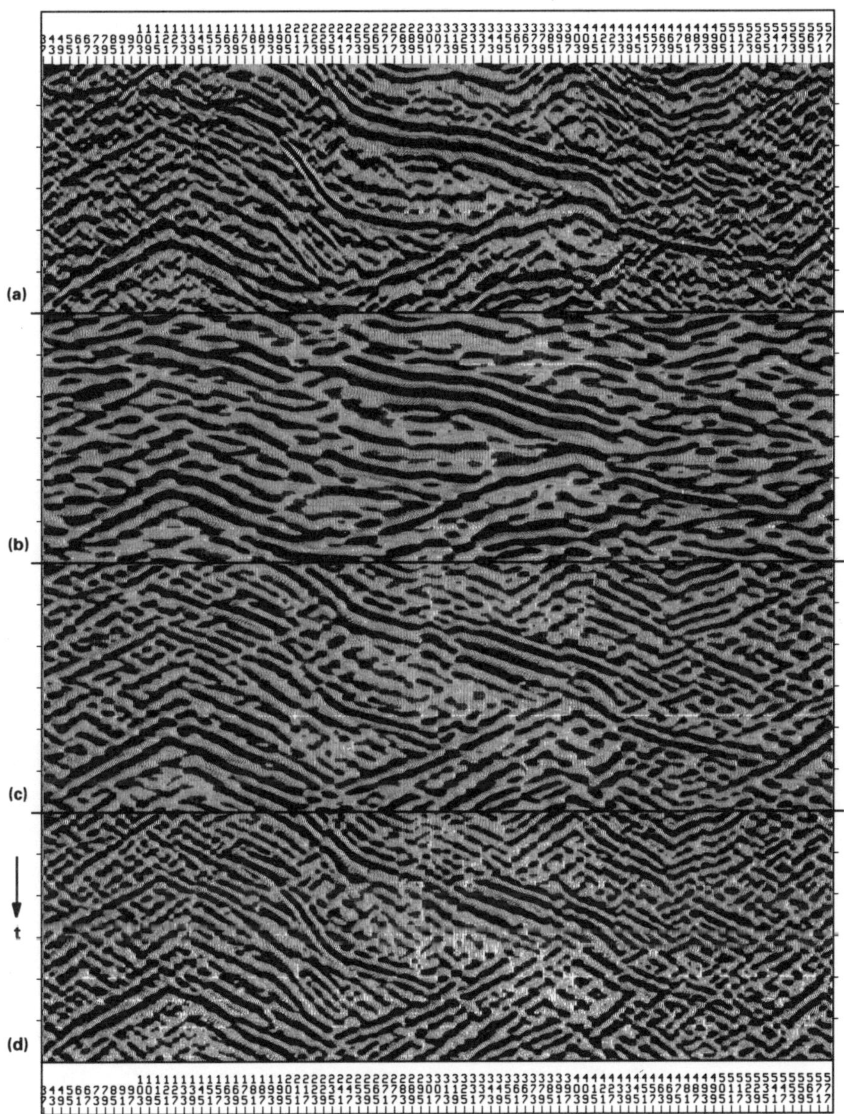

Fig. 2.3. Exclusive wavenumber filter test in analogy to exclusive frequency filter test. The corner points of the applied trapezium filters are expressed in percentage of k_N. (a) unfiltered 600 ms input window, (b) 0-0-8-12, (c) 2-6-12-20, (d) 4-8-20-32.

spatial antialias filter. Note that the very steep event from 1.8 to 4.0 s in the center of Figure 2.4a is now aliased, which it was not in the original data. Owing to the antialias filter the event is not present in Figure 2.4b. Also, on the right side of the section, below 2 s, many events that are incoherent in Figure 2.4a (representing energy around the Nyquist wavenumber k_N) have been removed by the antialias filter. If the steep events represent events to be interpreted we should not binate, of course.

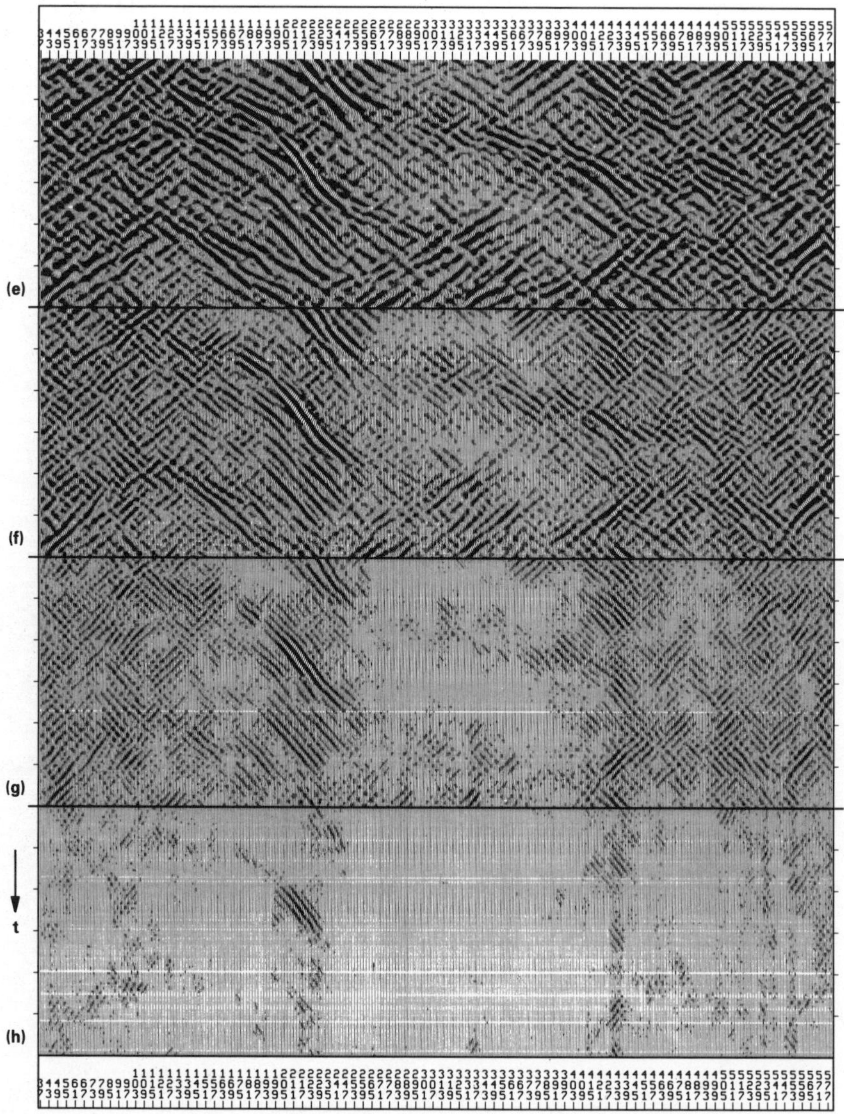

Fig. 2.3. cont. (e) 8-12-32-44, (f) 12-20-44-56, (g) 20-32-56-76, (h) 32-44-100-100.

In this chapter the two-dimensional seismic section is used to illustrate that:

—the usual seismic trace is just a collection of (t,x) points with constant x. Traces of (t,x) points with constant t can be treated similarly,

—sampling and resampling principles valid for temporal sampling are equally valid for spatial sampling, and

—zero-phase wavenumber filtering is entirely analogous to zero-phase frequency filtering.

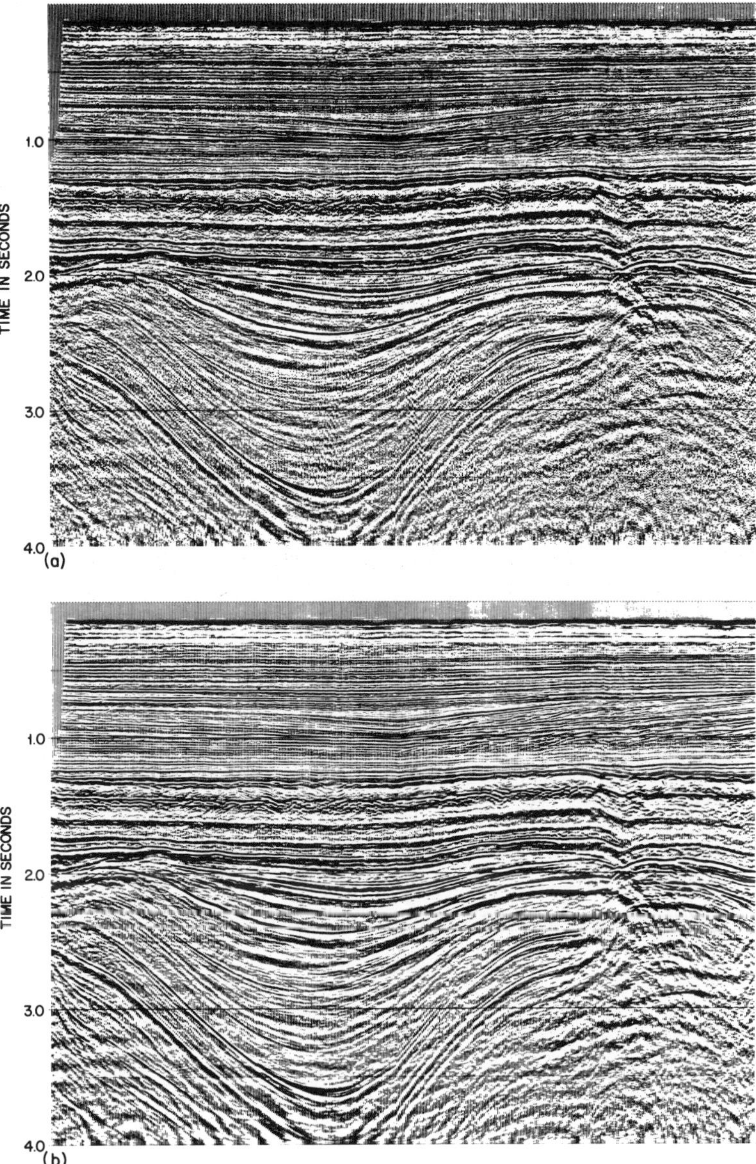

Fig. 2.4. Display showing every 6th trace of original data set (original data set not shown). (a) Without preceding spatial antialias filter, (b) After preceding spatial antialias filter. Note that the steeply dipping aliased events below 2.0 s in the center of (a) are not present in (b).

The discussion in this chapter forms the starting point for the extension in the next chapters to the three dimensions, one temporal and two spatial, that are necessary to describe the collection of shot records gathered for a 2-D multiple-coverage seismic line.

Chapter 3
THE CONTINUOUS WAVEFIELD

3.1 The Shot/Receiver- and Midpoint/Offset-Coordinate Systems

Chapter three mainly describes the properties of seismic data that can be recorded at (or close to) the surface. Elastic wave propagation through the earth is discussed only briefly.

Let us consider a point source (shot) at (x_s, y_s, z_s) emitting a unit spike at time $t = 0$. The displacement field of the point source can be recorded at any point (x_r, y_r, z_r) in three-dimensional (3-D) space. Hence, the response of this single physical experiment can be described by a function of t, x_r, y_r, and z_r. We assume that the same experiment carried out at some other physical time t' is repeatable, i.e., it will give rise to the same displacement field. The assumption of *repeatability* is valid only for a stationary earth and is not valid if the earth and hence the earth response would be different at different times, because of varying watertable, hydrocarbon production, steam injection, low and high tides, etc. The local nonelastic deformation caused by an explosive source also leads to nonrepeatability of the wavefield produced by shots from the same shot hole.

The shot can be placed anywhere in (x, y, z) and each time we get a different displacement field. We now define the *earth response* as the ensemble of displacement fields of all possible physical experiments with a unit point source, carried out at times when the experiments no longer influence each other. For all practical purposes we may assume that this earth response is a continuous function, i.e., two closely spaced shots would produce nearly identical displacement fields.

We strongly simplify our problem by considering only the component of the displacement field that is recorded by single-component geophones and we call this part of the earth response the *continuous wavefield W*. Thus

$$W = W(t, x_s, y_s, z_s, x_r, y_r, z_r) \tag{3.1}$$

is a scalar function of seven *independent* variables.

An abbreviated description of W can be given by introducing spatial vectors, the shot vector $\mathbf{x_s}(x_s, y_s, z_s)$ and the receiver vector $\mathbf{x_r}(x_r, y_r, z_r)$, i.e., $W = W(t, \mathbf{x_s}, \mathbf{x_r})$.

We can also describe the wavefield in terms of the midpoint vector $\mathbf{x_m}(x_m, y_m, z_m)$ and the offset vector $\mathbf{x_o}(x_o, y_o, z_o)$, i.e., $W = W(t, \mathbf{x_m}, \mathbf{x_o})$, in which $\mathbf{x_m}$ and $\mathbf{x_o}$ are defined by a linear coordinate transformation (Figure 3.1):

$$\mathbf{x_m} = (\mathbf{x_s} + \mathbf{x_r})/2,$$

and

$$\mathbf{x_o} = \mathbf{x_s} - \mathbf{x_r}. \tag{3.2}$$

We confine ourselves to the collection of multiple-coverage data gathered for a single seismic line and we assume the data to be gathered

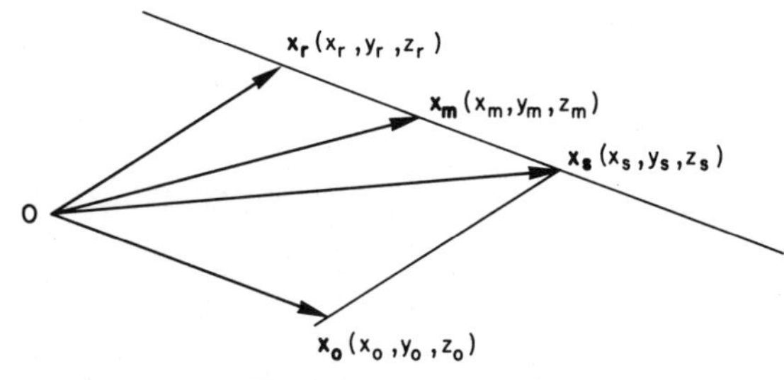

Fig. 3.1. The spatial vectors in physical space. $\mathbf{x}_s$ is shot vector, $\mathbf{x}_r$ is receiver vector, $\mathbf{x}_m$ is midpoint vector and $\mathbf{x}_o$ is offset vector.

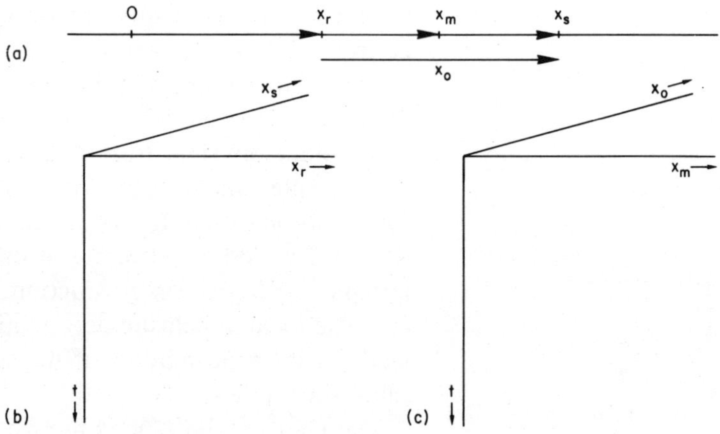

Fig. 3.2. The continuous wavefield along a seismic line expressed in two coordinate systems. (a) The four spatial coordinates in relation to the seismic line, (b) The shot/receiver coordinate system, (c) The midpoint/offset coordinate system.

along a straight line with constant depth d_s of the shots and constant depth d_r of the receivers. Some point on the seismic line is selected as the origin so that $y_s = y_r = 0$. Hence we need only to consider the subspace (t,x_s,x_r) and the continuous wavefield (consisting of infinitely many shots and infinitely many receivers for each shot) is a function of only three variables

$$W = W(t,x_s,x_r), \qquad (y_s = y_r = 0,\ z_s = d_s,\ z_r = d_r),$$

or

$$W = W(t,x_m,x_o).$$

Figure 3.2 illustrates the three-dimensional coordinate system; the usual convention is adopted of plotting increasing t downward.[1]

From equation (3.2) it follows that the three-dimensional shot/receiver coordinate system and the three-dimensional midpoint/offset coordinate system are linked by the simple linear transformations

---

[1]Similarly, the ensemble of vertical seismic profiles (VSPs) for a well can be described in the subspace (t,x_s,z_r) of the same seven-dimensional wavefield as specified in equation (3.1).

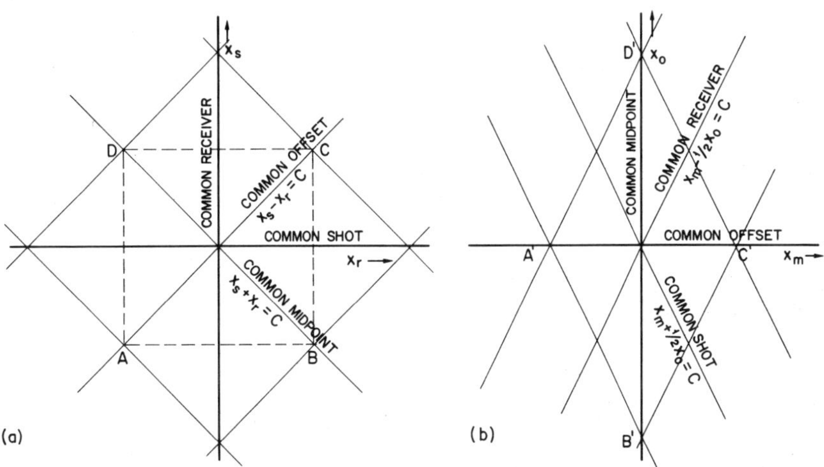

Fig. 3.3. Horizontal cross-section through three-dimensional coordinate systems ($t = t_c$). (a) The shot/receiver coordinate system with trajectories of constant (common) midpoint and offset, (b) The midpoint/offset coordinate system with trajectories of constant shot and receiver coordinates. The square ABCD in (x_s, x_r) is mapped onto A′B′C′D′ in (x_m, x_o). Note that area is invariant under the transformation.

$$x_m = (x_s + x_r)/2,$$
$$x_o = x_s - x_r,$$
$$t = t, \tag{3.3}$$

and

$$x_s = x_m + x_o/2,$$
$$x_r = x_m - x_o/2,$$
$$t = t. \tag{3.4}$$

Horizontal cross-sections through these coordinate systems are shown in Figure 3.3. It is instructive to check the mapping of square ABCD in (x_s, x_r) onto diamond A′B′C′D′ in (x_m, x_o) using equations (3.3) and (3.4). Note that all lines labeled "COMMON" in Figure 3.3 indicate direction only, not position. For instance, any line parallel to the x_r axis is a common shot trajectory with shot coordinate x_s.

A proper understanding of the behavior of various phenomena in both systems is essential for a proper choice of data acquisition and data processing parameters.

Many authors use half-offset rather than offset. Trajectories of constant shot and receiver coordinate in the midpoint/half-offset coordinate system are perpendicular to each other, whereas they cross obliquely in the midpoint/offset coordinate system (Figure 3.3). I prefer the "full-offset" transformation, and I use x_o rather than the more usual X for reasons of symmetry in the notation and to stress that x_o is a coordinate that can be positive and negative.

It may be noted that the Jacobian determinant of transformation (3.3)

$$
\begin{vmatrix} \dfrac{\partial x_m}{\partial x_s} & \dfrac{\partial x_m}{\partial x_r} \\[2ex] \dfrac{\partial x_o}{\partial x_s} & \dfrac{\partial x_o}{\partial x_r} \end{vmatrix}
=
\begin{vmatrix} \dfrac{1}{2} & \dfrac{1}{2} \\[2ex] 1 & -1 \end{vmatrix}
= -1.
$$

As a consequence area is invariant under the full-offset transformation, so that region ABCD in (x_s,x_r) maps onto an equal-area region A'B'C'D' in (x_m,x_o) (Figure 3.3).

3.2 Various Cross-Sections

Various cross-sectional panels (subspaces or domains) can be selected from the three dimensional (t,x_s,x_r) or (t,x_m,x_o) space (Figure 3.3). The common shot panel (CSP) is the (t,x_r) subspace. The wavefield in this cross-section can be determined by one physical experiment (one shot recorded at all possible receiver stations). In the other three spatial domains, common receiver panel (CRP), common midpoint panel (CMP) and common offset panel (COP), the wavefield must be deduced from a rearrangement of the ensemble of physical experiments. Another interesting cross-section is formed by the common time panel (CTP), that describes the wavefield for a constant value of time.

Throughout the following text the notations "CSP" and "(t,x_r) subspace" are used as equivalent expressions. The same holds for "CRP" versus "(t,x_s)", "COP" versus "(t,x_m)", and "CMP" versus "(t,x_o)". Note that in any common coordinate panel the other two coordinates vary. In the seismic literature the terms "profile," "gather," "record," or "slice" are often used as an equivalent to "panel".

A special cross-section is the zero-offset panel for which source and receiver coincide. The zero-offset panel serves as the symmetry plane for reciprocity in (t,x_m,x_o), as shown later. In many approaches to seismic processing the zero-offset panel for one wave type only (compressional or shear) serves as an intermediate stage to the final migrated section. Then the nonzero-offset panels are used to remove unwanted wave types and noise.

The common time panel is the least familiar of the possible cross-sections. CTPs or (x_s,x_r) panels can be constructed from the three-dimensional prestack seismic data in the same way as time slices from a 3-D seismic survey. However, they show very different phenomena. Section 3.9 gives some examples of wavefields in this cross-section.

Without further introduction cross-sections through various transformed representations of $W(t,x_s,x_r)$ will also be used in this book, e.g., common frequency panel in Figure 3.9.

3.3 Apparent Velocity in Various Subspaces

The properties of the rather abstract continuous wavefield can be derived using the more familiar concept of *event* or *lineup*. An event is formed by coherent seismic energy corresponding to one of the wave types that have traveled from source to receiver via some path through the subsurface. Together the events constitute the continuous wavefield. Examples of events are primary and multiple reflections, converted waves, groundroll or surface waves (usually consisting of many lineups), and diffractions. The arrival time of an event forms a surface (the traveltime surface) in the three-dimensional coordinate systems (t,x_s,x_r) and (t,x_m,x_o).

An important property of elastic wave propagation is that the arrival time t of each event is a continuous function of the two spatial coordinates, $t = t(x_s,x_r) = t(x_m,x_o)$ and we may consider the *partial derivatives* of this function with respect to each coordinate. Using equation (3.4) and the chain rule we can express the partial derivatives in one coordinate system in the partial derivatives in the other system:

$$\frac{\partial t}{\partial x_m} = \frac{\partial t}{\partial x_s}\frac{\partial x_s}{\partial x_m} + \frac{\partial t}{\partial x_r}\frac{\partial x_r}{\partial x_m} = \frac{\partial t}{\partial x_s} + \frac{\partial t}{\partial x_r},$$

and

$$\frac{\partial t}{\partial x_o} = \frac{\partial t}{\partial x_s}\frac{\partial x_s}{\partial x_o} + \frac{\partial t}{\partial x_r}\frac{\partial x_r}{\partial x_o} = \frac{1}{2}\frac{\partial t}{\partial x_s} - \frac{1}{2}\frac{\partial t}{\partial x_r}.$$

Similar relations are derived and used in Claerbout (1985) for the shot/receiver and midpoint/halfoffset coordinate systems. The spatial derivatives of the travel time surface of an event have the dimension of slowness or inverse velocity. For reasons to be given later in this section we call $\partial t/\partial x_i$ the *apparent slowness* p_i of the event in the cross-section defined by (t,x_i). Thus we have found a simple relationship between the apparent slownesses in the various cross-sections:

$$p_m = p_s + p_r,$$

and

$$p_o = (p_s - p_r)/2. \tag{3.5}$$

Likewise

$$p_s = p_m/2 + p_o,$$

and

$$p_r = p_m/2 - p_o. \tag{3.6}$$

Note that all slownesses are functions of (x_s,x_r) or (x_m,x_o).

In ray theory travel paths of rays traveling through the subsurface are studied. There it is customary to call $\partial t/\partial x$ the horizontal slowness p and $\partial t/\partial z$ the vertical slowness q. Our slowness p_r corresponds to the horizontal slowness as "seen" by the line of receivers and is the inverse of the *apparent* velocity $V_r(x_s = C,x_r)$ at which the event travels along the receiver positions. As a naming convention we call all slownesses and velocities in the various panels apparent slownesses and apparent velocities, (even though in the other panels the event is not a physically traveling wave).

Note that equations (3.5) and (3.6) are valid for any continuous and differentiable surface in (x_s, x_r). The validity is not restricted to the arrival times of an event. The equations would also apply to the maximum of the envelope of an event (describing the apparent *group* velocity of the event) and to the surface of constant phase of a frequency component of an event (describing the apparent *phase* velocity of the event).

The expressions for the apparent velocities corresponding to equations (3.5) and (3.6) are:

$$\frac{1}{V_m} = \frac{1}{V_s} + \frac{1}{V_r},$$

$$\frac{1}{V_o} = \frac{1}{2}\left(\frac{1}{V_s} - \frac{1}{V_r}\right), \tag{3.7}$$

and

$$\frac{1}{V_s} = \frac{1}{2V_m} + \frac{1}{V_o},$$

$$\frac{1}{V_r} = \frac{1}{2V_m} - \frac{1}{V_o}. \tag{3.8}$$

These relations illustrate the three-dimensional nature of each single event. At any point (x_s, x_r) or (x_m, x_o) an event has an apparent velocity V_r in the CSP through that point and another apparent velocity V_s in the CRP through that point. If we know V_s and V_r, then V_m and V_o can be readily computed.

The application of the normal moveout (NMO) correction to a reflection event converts this event in each CMP to a horizontal event for which $1/V_o = 0$. Then it follows from equation (3.7) that

$$\frac{1}{V_s} = \frac{1}{V_r} \quad \text{and} \quad \frac{1}{V_m} = \frac{2}{V_s}.$$

In an NMO-corrected data set the apparent velocity of a reflection event in a COP is shown to be half the apparent velocity in the corresponding CRPs and CSPs.

Section 3.8 provides more examples of the use of formulas (3.3) through (3.8).

Before continuing with the notion of *minimum* apparent velocity, we must first consider the reciprocity theorem.

3.4 The Reciprocity Theorem

Numerous authors have written about the reciprocity theorem. Seismic reciprocity, i.e., reciprocity applicable to media and disturbances for which the *elastic* wave equation is valid, was proved in Knopoff and Gangi (1959). White (1960) reformulated the theorem in a lucid "Statement of reciprocity". A more recent generalization of the theorem can be found in Bojarski (1983).

For the purpose of this text White's formulation is worth quoting:

STATEMENT OF RECIPROCITY

"If, in a bounded inhomogeneous, anisotropic, elastic medium, a transient force $f(t)$ applied in some particular direction α at some point P creates at a second point Q a transient displacement whose component in some direction β is $u(t)$, then the application of the same force $f(t)$ at point Q in the direction β will cause a displacement at point P whose component in the direction α is $u(t)$.

Note that the statement applies to the whole disturbance, whether it be body waves, surface waves, or whatever. It is not restricted to 'far-field' geometry."

Despite the wide applicability of the reciprocity theorem, some important restrictions are worth emphasizing. First, the medium must behave elastically, hence it must return to its original position; forces that generate a permanent deformation do not obey reciprocity. Second, the theorem applies to equal directional input forces resulting in equal directional displacements. In seismic experiments it may be difficult or even impossible to generate the same directional force at a pair of locations P and Q with different elastic properties at the surface. Third, source and receiver must exchange position. A configuration with a buried source and a receiver at the surface is reciprocal with a surface source and a buried receiver. In seismic data acquisition sources and receivers are usually at a constant depth d_s, and d_r, respectively, to which reciprocity does not apply if $d_s \neq d_r$.

In marine data acquisition the scalar pressure field is recorded, there reciprocity applies without the directionality restriction.

The reciprocity theorem is an assertion about a single experiment carried out twice with an exchange of source and receiver positions, i.e.

$$W(t,x_s = x_p, x_r = x_q) = W(t,x_s = x_q, x_r = x_p)$$

in which x_p and x_q describe the coordinates of points P and Q in (x,y,z).

For reciprocity to be valid for the *ensemble* of all shots along the seismic line it is necessary that the directions α and β in the "Statement of reciprocity" are equal, and that the depths d_s and d_r are equal. Assuming $\alpha = \beta$ (for land data acquisition) and $d_s = d_r$, then reciprocity means for the continuous wavefield that

$$W(t,x_s,x_r) = W(t,x_r,x_s), \tag{3.9}$$

and

$$W(t,x_m,x_o) = W(t,x_m,-x_o). \tag{3.10}$$

Equation (3.10) shows that $W(t,x_m,x_o)$ is symmetric with respect to $x_o = 0$, being an even function of x_o, which means that all odd derivatives equal zero at $x_o = 0$. In particular

$$\frac{\partial W(t,x_m,x_o)}{\partial x_o} = 0 \text{ for } x_o = 0, \tag{3.11}$$

and also

$$\frac{\partial t(x_m,x_o)}{\partial x_o} = 0 \text{ for } x_o = 0, \qquad (3.12)$$

in which $t(x_m,x_o)$ describes the traveltime surface of an event in (t,x_m,x_o).

Another consequence of the reciprocity theorem is that the wavefield generated by a source at P for receivers along x equals the wavefield recorded at P for (individual) sources along x, i.e.:

$$W(t,x_s = x_p,x) = W(t,x,x_r = x_p), \text{ for all } x. \qquad (3.13)$$

The reciprocity theorem holds for the whole wavefield, including the "shot-generated" noise (as long as it can be described by the *elastic* wave equation). Therefore, shot-generated is a somewhat misleading term: though the source takes care of the energy and the frequency spectrum of the source wavelet, the noise is part of the earth response and is dependent on the properties of the medium.

Reciprocity does not reduce the three-dimensional wavefield to a two-dimensional wavefield. Equation (3.9) shows that reciprocity is a *symmetry* relation which does not reduce the number of independent coordinates.

In off-end shooting, the wavefield is recorded only for positive or negative values of the x_o coordinate, hence, off-end shooting assumes that reciprocity applies. As a consequence the seismic lines may be shot in either direction, and parallel lines may be shot in opposite directions.

Reciprocity also applies to data shot and recorded with shot and receiver patterns provided the shot patterns are identical to the receiver patterns (Symmetric sampling, Section 4.6.8).

In practice deviations from the theoretical requirements occur frequently but usually do little harm. Reciprocity does not apply if depths of shots and receivers are constant but different, $d_s \neq d_r$, yet in that case there will not be much difference between the far-field parts of $W(t,x_s,x_r)$ and $W(t,x_r,x_s)$. However surface waves and near-surface effects, such as statics could be very different.

Reciprocity is achieved easiest in land data acquisition if a directional surface source such as Vibroseis is used in combination with geophones planted at the surface. Then reciprocity does not only apply to vertical P-wave recording but also to horizontal S-wave recording provided the same direction is used for shot and receiver. Reciprocity would not apply for a cross-line source and an in-line receiver or vice versa.

In PSV recording the source may generate a vertical P-wave whereas reflected S-waves are recorded with horizontal geophones. Obviously reciprocity does not apply to these data. Interesting experiments can be carried out using various combinations of directional forces and displacements.

3.5 Minimum Apparent Velocity

Let us now examine the notion of minimum apparent velocity. This velocity will play an important role in the description of the energy distribution in the frequency/wavenumber spectra of the wavefield.

Theoretically, the smallest layer velocity that can occur is zero. This follows from the expression for the shear wave velocity V_S in an isotropic medium with shear modulus μ and density ρ:

$$V_S = \sqrt{\frac{\mu}{\rho}}.$$

In fluids $\mu = 0$, with no propagation of shear wave energy. In marshlands and other unconsolidated media V_S can be very small. Likewise, the Rayleigh wave velocity V_R is very small in those media ($V_R < V_S$). However, in most practical cases there is a lower limit, larger than zero, to the layer velocities in the area of investigation.

For these discussions our interest is in the *phase velocity* of the disturbance. Phase velocity is the velocity with which the wavefront (a surface of constant phase) of a disturbance or event propagates through the medium. The phase velocity may be frequency dependent (in particular for groundroll) and does depend in a complicated way on the elastic properties of the medium. It may be assumed there is a lower limit, larger than zero, to the phase velocities, if there is one for the layer velocities. I call this lower limit of the phase velocity (in the area of investigation) the minimum phase velocity $V_{\min}$. However, bear in mind that in some areas $V_{\min}$ can be very small.

In the CSP (the only panel representing a physical experiment) the apparent velocity V_r of an event depends on the (phase) velocity V of the wavefront at the receivers and on the angle α between the wavefront and the line of receivers (Figure 3.4):

$$V_r = V/\sin\alpha.$$

V_r is smallest for horizontally traveling waves such as groundroll and direct arrival, then $\alpha = 90°$ and $V_r = V$. Thus there exists a minimum apparent velocity $|V_r|_{\min}$ that equals the smallest of all V's along the line of receivers, i.e., the steepest possible event has an apparent velocity

$$|V_r|_{\min} = V_{\min}.$$

Note that the above reasoning is also valid if the normal to the wavefront does not lie in the vertical plane. Events coming in from the side will also always have an apparent velocity greater than or equal to $V_{\min}$.

The reciprocity theorem as expressed in equation (3.13) means that

$$|V_s|_{\min} = |V_r|_{\min} = V_{\min}.$$

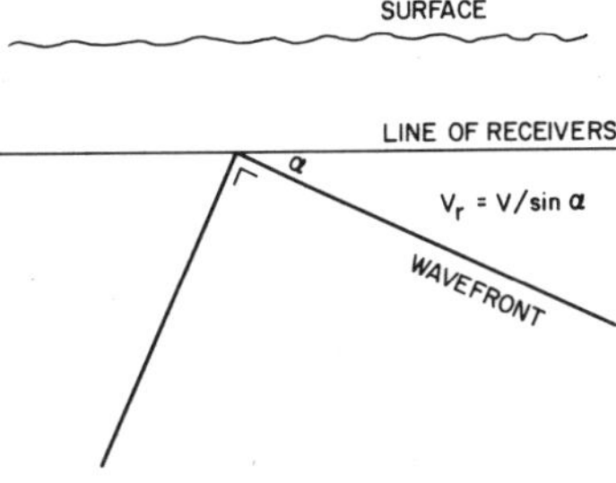

Fig. 3.4. Wavefront of an event reaching the line of receivers. V_r is the apparent velocity at which the wave travels along the line of receivers. V is the phase velocity of the event.

Minimum apparent velocities also exist in CMP and COP. Inspection of equation (3.7) shows that the maximum for $1/|V_m|$ is reached if V_s and V_r are both equal to $+V_{min}$ or $-V_{min}$, hence

$$|V_m|_{min} = \frac{1}{2} V_{min},$$

whereas the maximum for $1/|V_o|$ is reached if $|V_s| = |V_r| = V_{min}$ and V_s, V_r have opposite sign, hence

$$|V_o|_{min} = V_{min}.$$

Summarizing, we have

$$|V_s|_{min} = |V_r|_{min} = |V_o|_{min} = V_{min},$$

and

$$|V_m|_{min} = \frac{1}{2} V_{min}. \tag{3.14}$$

The direct arrival and back scatter can be used to illustrate that $|V_o|_{min}$ and $|V_m|_{min}$ can indeed occur (Sections 3.8.1 and 3.8.2).

3.6 Transforms of the Continuous Wavefield

3.6.1 Wavenumber transforms

Chapter 2 mentions the spatial spectrum of a two-dimensional seismic section. We now introduce the spatial spectra of the three-dimensional continuous wavefield. Here we can distinguish four wavenumbers k_i, one for each spatial coordinate x_i, $i = s, r, m, o$. For instance,

$$W_r(t,x_s,k_r) = \int_{-\infty}^{\infty} W(t,x_s,x_r) \exp(2\pi j k_r x_r) dx_r$$

defines the wavenumber k_r corresponding to x_r. The subscript r in W_r indicates integration over x_r. In this definition k_r is the inverse of apparent wavelength in analogy to frequency f as the inverse of period. (Other authors define k in analogy to radial frequency ω.) Likewise, k_s, k_m and k_o can be defined. The wavenumber spectrum $W_r(t,x_s,k_r)$ depends on the parameters x_s and t.

Strictly speaking, we should call k_s, k_r, k_m and k_o *apparent* wavenumbers, as we measure the wavefield variations along the horizontal through shots and receivers, as opposed to measurement along the travel paths. For convenience we use "wavenumber" for "apparent wavenumber" in the sequel.

Note that k_s and k_r are *independent* variables, and should be carefully distinguished from each other. Using k for prestack seismic data without using one of the subscripts s, r, m, or o can be confusing.

A double Fourier transform of any (t,x_i) panel leads to a (f,k_i) spectrum, e.g. $W_{ts}(f,k_s,x_r)$.

A triple Fourier transform of the whole continuous wavefield leads to the three-dimensional (f,k_s,k_r) spectrum $W_{tsr}(f,k_s,k_r)$, or the (f,k_m,k_o) spectrum $W_{tmo}(f,k_m,k_o)$.

The coordinate transformations defined by equations (3.3) and (3.4) can easily be shown to correspond to coordinate transformations $(k_s, k_r) \rightarrow (k_m, k_o)$ given by

$$k_m = k_s + k_r,$$
$$k_o = (k_s - k_r)/2, \tag{3.15}$$

and

$$k_s = k_m/2 + k_o,$$
$$k_r = k_m/2 - k_o. \tag{3.16}$$

Note the similarity of these expressions to equations (3.5) and (3.6). The wavenumber coordinate systems are illustrated in Figure 3.5. The Jacobian determinants of transformations (3.15) and (3.16) are equal to -1. Hence, similar to the transformation of (x_s, x_r) to (x_m, x_o) (Section 3.1), area is invariant under the transformations.

Reciprocity leads to symmetry relations in the double wavenumber spectra analogous to equations (3.9) and (3.10):

$$W_{tsr} (f, k_s, k_r) = W_{tsr} (f, k_r, k_s), \tag{3.17}$$

and

$$W_{tmo} (f, k_m, k_o) = W_{tmo} (f, k_m, -k_o). \tag{3.18}$$

Here again the subscripts of W indicate the integration variables.

Wavenumber filtering is entirely analogous to frequency filtering, as discussed in Chapter 2. It is important to identify clearly in which domain the wavenumber filter is applied. For instance, the application of a shot domain filter $h(x_s)$ (in the field this would be called a shot pattern) does not lead to the same results as the application of a receiver domain filter $h(x_r)$ (receiver pattern), even if the filter is the same, i.e., $W(t, x_s, x_r) * h(x_s) \neq W(t, x_s, x_r) * h(x_r)$ (Figure 3.6).

The continuous wavefield is no longer symmetrical with respect to $x_s = x_r$ (zero offset, $x_o = 0$) after the application of a single wavenumber filter. Only if a two-dimensional symmetrical wavenumber filter is applied, e.g., $h(x_s) * h(x_r)$, will the symmetry relation survive.

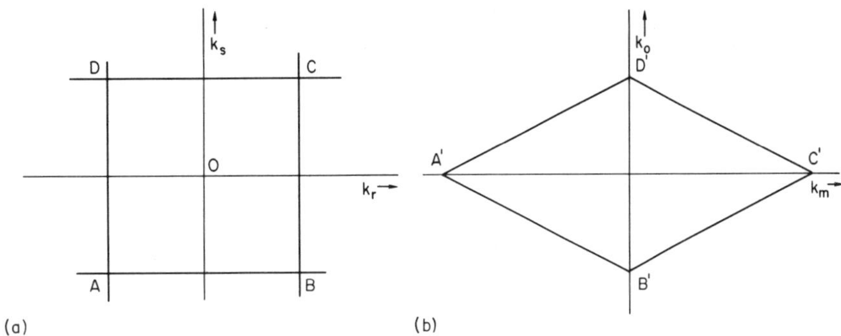

Fig. 3.5. Double wavenumber coordinate systems. (a) Shot/receiver wavenumber coordinate system, (b) Midpoint/offset wavenumber coordinate system. The square ABCD in (k_s, k_r) is mapped onto A'B'C'D' in (k_m, k_o). Note that area is invariant under the transformation.

Frequency f and (apparent) wavenumber k_i of a monochromatic plane wave traveling at apparent velocity V_i along the x_i axis are related by

$$k_i = f/V_i, \qquad i = s, r, m, o. \tag{3.19}$$

This expression is a generalization of the well-known formula $k = f/V$ or $\lambda = VT$ with λ is wavelength and T is period. Equation (3.19) defines a straight line in (f,k_i) for a nondispersive event traveling at apparent velocity V_i (Figure 3.7).

3.6.2 (τ, p) transforms

The transformation to the domain of intercept time τ and apparent slowness p has gained some interest in the 1980s. Though I will not discuss its applications in any detail, within the context of this book I do note that four different (τ, p) transforms can be distinguished. For the CSP the (τ, p) transform is defined as

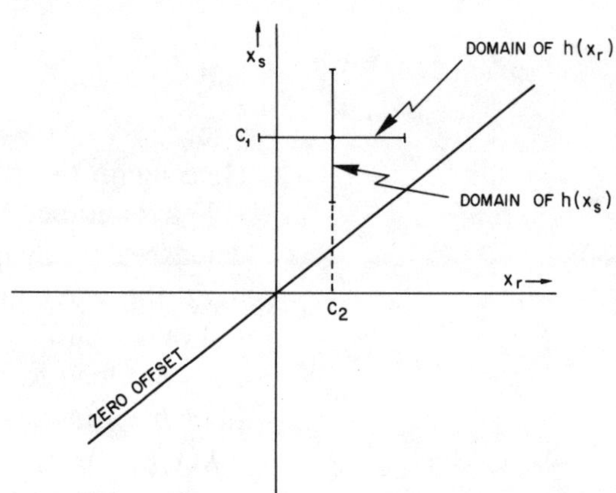

Fig. 3.6. Inequality of wavenumber filtering in various domains. Application of $h(x_r)$ would filter points along $x_s = C_1$, whereas application of $h(x_s)$ would filter points along $x_r = C_2$, leading to different output values.

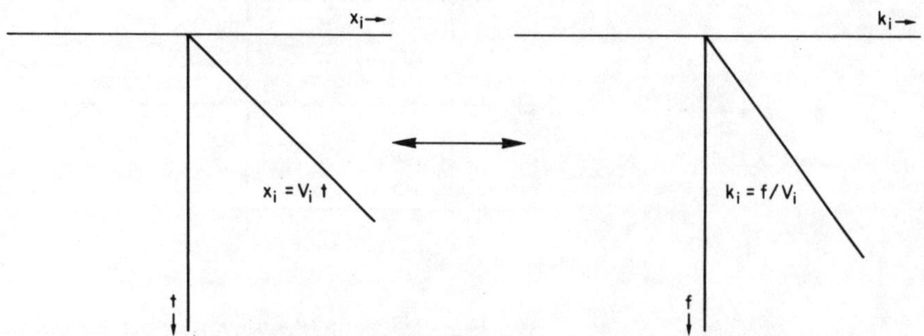

Fig. 3.7. A plane wave of apparent velocity V_i in (t,x_i) corresponds to a straight line $k_i = f/V_i$ in (f,k_i).

$$\Phi_r(\tau,x_s,p_r) = \int_{-\infty}^{\infty} W(\tau + p_r x_r, x_s, x_r)dx_r.$$

A value of p_r represents lines of constant apparent slowness or dip in the CSP. Similar relations can also be written for the CRP, CMP, and COP. The (τ,p_o) transform maps a hyperbola in (t,x_o) onto an ellipse in (τ,p_o) (cf. Diebold and Stoffa, 1981, Claerbout, 1985).

A double (τ,p) transform can also be defined as

$$\Psi(\tau,p_s,p_r) = \int_{-\infty}^{\infty} \int_{-\infty}^{\infty} W(\tau + p_s x_s + p_r x_r, x_s, x_r)dx_s dx_r.$$

This transform would lead to traces in the three-dimensional (τ,p_s,p_r) space. Each point $\Psi(\tau,p_s,p_r)$ corresponds to the stack of all data in a plane defined by

$$t = \tau + p_s x_s + p_r x_r.$$

$\Psi(\tau,0,0)$ would be the horizontal stack of the complete wavefield.

Applications of the single (τ,p) transform are described in Diebold and Stoffa, 1981, Stoffa et al, 1981, Schultz, 1982, Ottolini and Claerbout, 1984, and many other authors. Applications of the double (τ,p) transform are yet to be published.

3.7 The Energy Distribution of the Continuous Wavefield

Investigating the energy distribution of the continuous wavefield in (f,k_s,k_r) is of interest.

From equations (3.14) and (3.19) it follows that in the (f,k_s) and (f,k_r) panels the lines

$$k_{s,r} = \pm f/V_{min}$$

divide the panels into regions with and without energy (Figure 3.8a). (The notation $k_{s,r}$ is shorthand for k_s or k_r.)

For a given frequency f the wavenumber spectrum is confined to

$$-f/V_{min} \leq k_{s,r} \leq f/V_{min}.$$

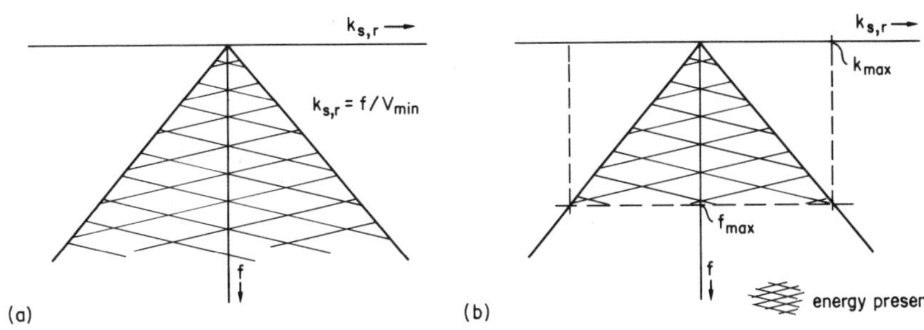

Fig. 3.8. Regions with and without energy in (f,k_s) and (f,k_r). (a) Minimum apparent velocity determines region in which energy is present, (b) If there is a maximum frequency, then there is also a maximum wavenumber.

This property is also referred to as *spatial bandwidth limitation* (Berkhout, 1984). In the cross-section (k_s, k_r) or common frequency panel this leads to a square as in Figure 3.9a. In the three-dimensional (f, k_s, k_r) space energy is confined to a pyramid shaped volume (Figure 3.10).

Following the same reasoning for the (f, k_m, k_o) space the energy in a (k_m, k_o) cross-section is found to be confined by lines

$$k_m = \pm 2f/V_{\min} \quad \text{and} \quad k_o = \pm f/V_{\min},$$

according to equations (3.14) and (3.19). However, mapping the region with energy in (k_s, k_r) on (k_m, k_o) shows that the energy in (k_m, k_o) is restricted to the diamond-shaped area in Figure 3.9b.

The difference between energy regions in (k_m, k_o) and $k_s, k_r)$ can also be illustrated with direct arrivals and back scatter (Sections 3.8.1 and 3.8.2 deal with these events in more detail).

Direct arrivals traveling at the minimum velocity $V_{\min}$ have energy

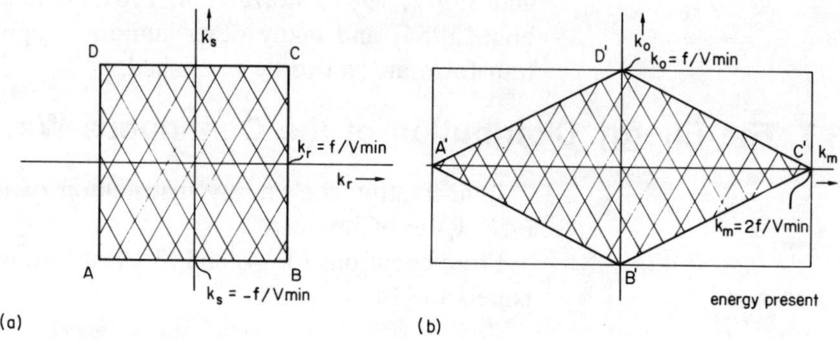

Fig. 3.9. Regions with and without energy in common frequency panel. (a) Energy in (k_s, k_r) is confined by lines $k_{s,r} = \pm f/V_{\min}$. (b) Energy in (k_m, k_o) is not confined by lines $k_m = \pm 2 f/V_{\min}$ and $k_o = f/V_{\min}$ but by diamond-shaped mapping of cross-hatched area in (a).

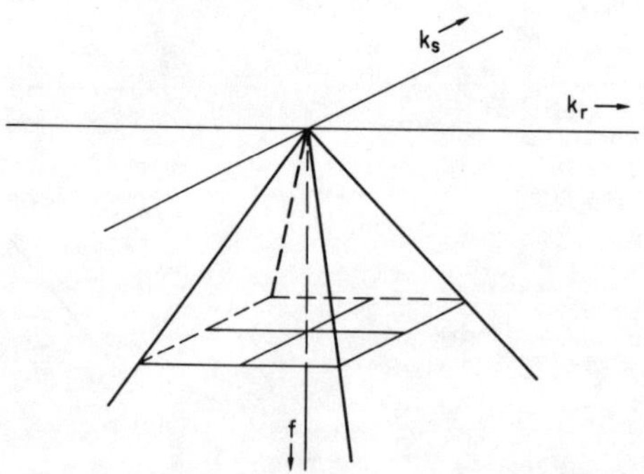

Fig. 3.10. Energy in (f, k_s, k_r) is confined to a pyramid-shaped volume. If there exists a maximum frequency $f_{\max}$, the base of the pyramid is given by the lines $k_{s,r} = \pm f_{\max}/V_{\min}$.

only at $k_m = 0$, $k_o = \pm f/V_{\min}$, as follows from equation (3.21b). (Points B′ and D′ corresponding to points B and D in Figure 3.9.)

Back scatter traveling at the minimum velocity $V_{\min}$ has energy only at $k_m = \pm 2f/V_{\min}$, $k_o = 0$, as follows from equation (3.22d). (Points A′ and C′ corresponding to points A and C in Figure 3.9.)

Hence, direct arrivals and back scatter illustrate that the continuous wavefield may have energy in all corners of the square ABCD in Figure 3.9a.

If $W_{tsr}(f,k_s,k_r) = 0$ for $f > f_{\max}$, then there exist *maximum wavenumbers* $|k_{s,r}|_{\max} = f_{\max}/V_{\min}$, see Figure 3.8b.

The pyramid of Figure 3.10 is then bounded at the base by lines $k_{s,r} = \pm f_{\max}/V_{\min}$. Similar reasoning applies to (k_m,k_o). Thus:

$$|k_s|_{\max} = |k_r|_{\max} = |k_o|_{\max} = f_{\max}/V_{\min},$$

and

$$|k_m|_{\max} = 2f_{\max}/V_{\min}. \tag{3.20}$$

So far I have discussed regions to which the energy of the wavefield is confined. The energy distribution within these regions is determined by the elastic properties of the subsurface and its structural configuration.

Horizontal events in any vertical cross-section have $V_i = \infty\,(i = s,r,m,o)$, so that according to equation (3.19) $k_i = 0$. Then there is no energy for $|k_i| > 0$. In a flat homogeneous and isotropic earth COPs consist of horizontal events only, then COPs have energy only for $k_m = 0$, whereas in the CMP energy is present for a range of k_o values due to the moveout of the data (Figure 3.11a). The more dipping events, faults and diffraction points occur, the more energy will be found for the larger k_m values. A conceptual energy distribution in (k_m,k_o) for a normal geology is illustrated in Figure 3.11b. This geology has no preference for right or left dipping events. If there exist only positive dips (traveltime increasing with increasing x_m), and if there are no diffractions, then there is only energy in (k_m,k_o) for $k_m \geq 0$.

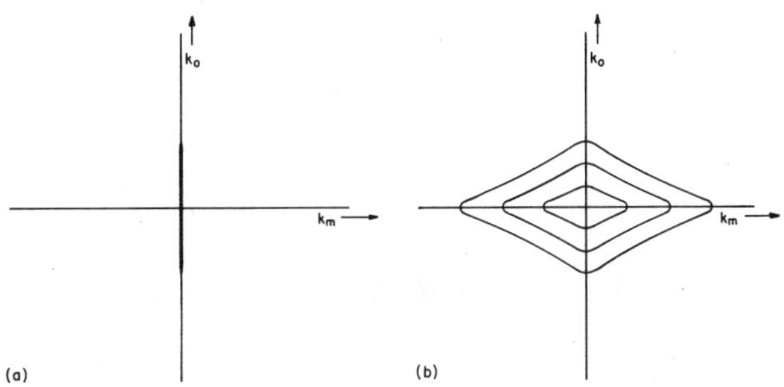

Fig. 3.11. Equal energy contours in (k_m,k_o). (a) Horizontal layers only. (b) Normal geology. Symmetry with respect to $k_o = 0$ follows from reciprocity principle. Symmetry with respect to $k_m = 0$ indicates no preference for right or left dips.

The energy distribution in (k_m, k_o) determines the compromises that can be accepted in the selection of data acquisition parameters. Much energy for large k_m values requires small shot and receiver sampling intervals. Sampling is discussed in Chapter 4.

3.8 Examples of Events in (t, x_s, x_r) and (t, x_m, x_o)

It is highly instructive to study the behavior of various events in (t, x_s, x_r) or (t, x_m, x_o). Each event in a cross-section can be represented by a traveltime surface in (t, x_m, x_o). In fact the analysis of the shape of events in (t, x_m, x_o) may lead to better discrimination between the various wave types, e.g. the typical shape of a diffraction may be distinguished from the shape of dipping events. This may prevent misinterpretation of the final processed data.

This section looks at the direct wave, a diffraction point on the line, a diffraction point at some distance from the line, and dipping events.

3.8.1 The direct wave

The direct wave is an event that travels in a straight line from source to receiver. In a medium with constant velocity V its traveltime is given by:

$$t(x_s, x_r) = |x_s - x_r|/V \tag{3.21a}$$

The corresponding expression in (x_m, x_o) reads [using equation (3.4)]:

$$t(x_m, x_o) = |x_o|/V. \tag{3.21b}$$

Hence, if V is constant along the line, the traveltime is a function only of x_o and not of x_m. This is illustrated in Figure 3.12, which shows various cross-sections through the direct wave. A three-dimensional representation of the event shows its tent shape (Figure 3.14), and illustrates that the direct wave is a two-dimensional event having the same shape in any CSP, CRP, or CMP.

The apparent velocities in the various domains follow from

$$\frac{1}{V_i} = \frac{\partial t}{\partial x_i}, \qquad \text{and} \quad \text{equation (3.21):}$$

or

$$1/V_s = 1/V_r = 1/V_o = \pm 1/V, \qquad \text{and} \quad 1/V_m = 0.$$

This equation shows that the direct wave is an event with constant apparent velocity $\pm V$ in CSP, CRP, and CMP, whereas it is a horizontal event in the COP.

3.8.2 A diffraction point on the seismic line

In land seismic data acquisition a local disturbance in surface conditions such as a ditch or dike crossing the seismic line will act as a diffraction point at zero distance from the seismic line. (The general case of a diffraction point at some distance from the seismic line is treated in the next section.)

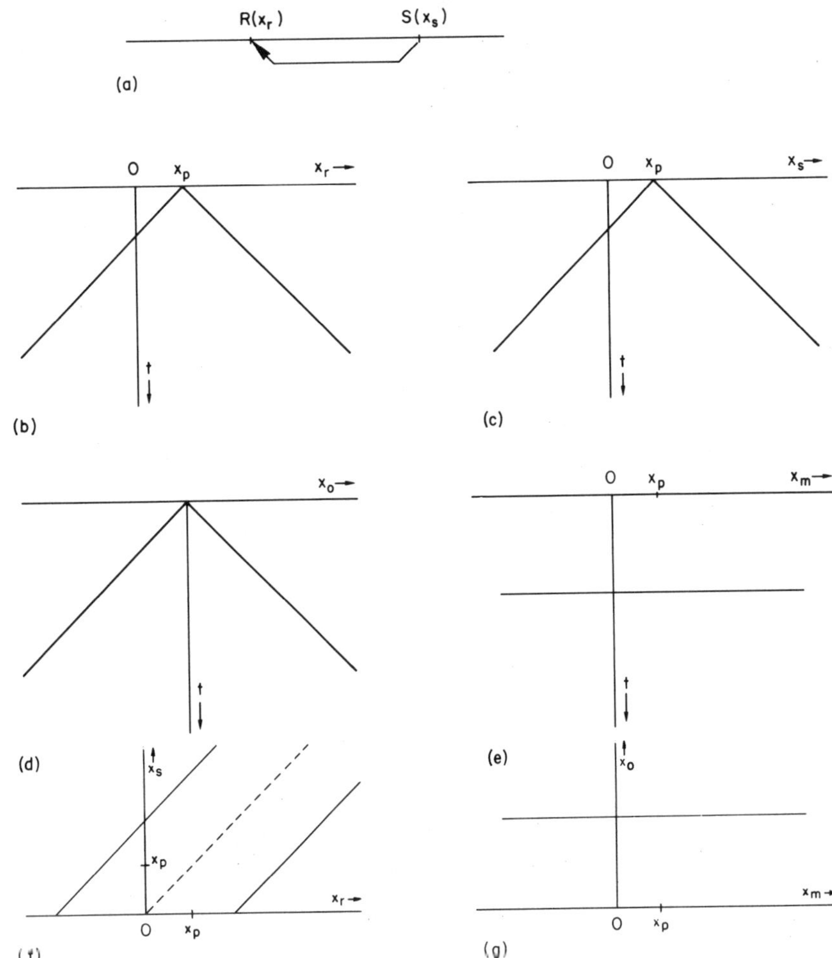

Fig. 3.12. Continuous wavefield of event traveling directly from shot to receiver in a constant-velocity medium. (a) geometry, (b) CSP for $x_s = x_p$, (c) CRP for $x_r = x_p$, (d) CMP for $x_m = x_p$, (e) COP for $x_o = C$, (f) CTP in (x_s,x_r) for $t = C/V$, (g) CTP in (x_m,x_o) for $t = C/V$. In (t,x_s,x_r) this event has the shape of an infinitely long tent.

The geometry of this event is described in Figure 3.13a,b,c. The origin is selected at the diffraction point D. A distinction must be made between the cases in which shot and receiver are on opposite sides of D (case 1, Figure 3.13b) or on the same side (case 2, Figure 3.13c). In case 1 the event is no different from the direct wave described in the previous section. In case 2 the wave is also known as back scatter. Back-scattered groundroll can be observed in Figure 4.16.

Thus the traveltime surface is described in (x_s,x_r) by:

$$t(x_s,x_r) = |x_s - x_r|/V \qquad \text{for case 1,} \qquad (3.22a)$$

and

$$t(x_s,x_r) = |x_s + x_r|/V \qquad \text{for case 2.} \qquad (3.22b)$$

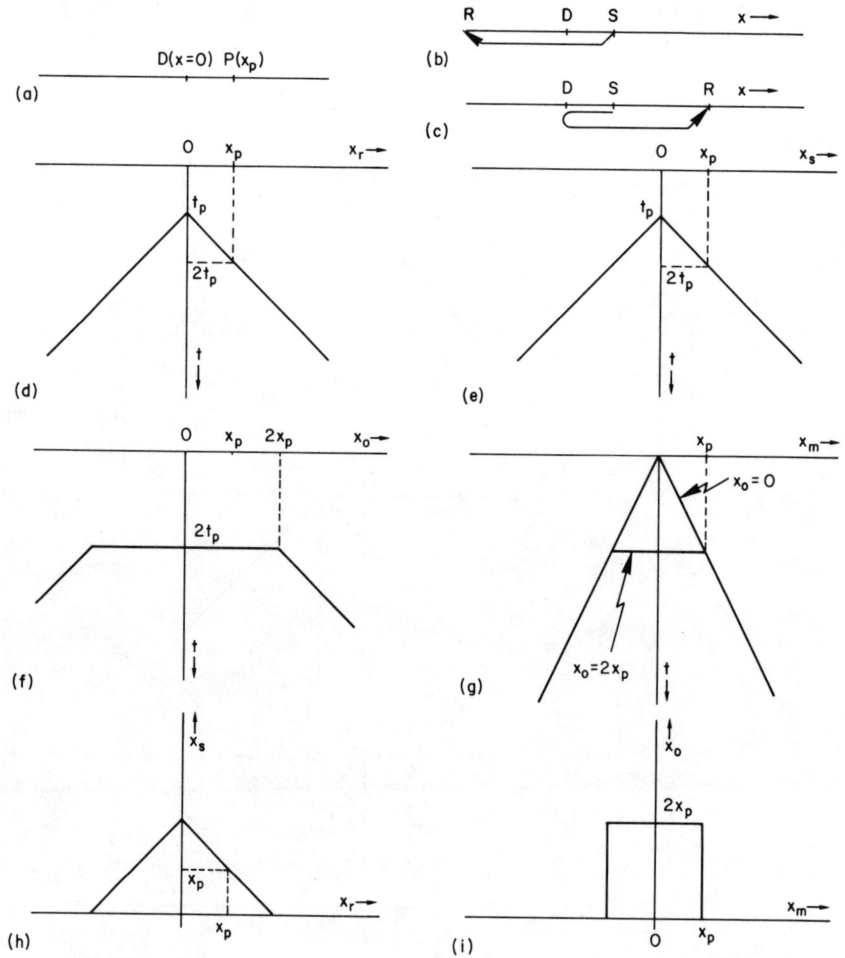

Fig. 3.13. Continuous wavefield generated by diffraction point situated on the seismic line. (a) Geometry, diffraction point D at $x = 0$, P is some point at $x = x_p$, (b) Diffraction point between shot and receiver, in this case the traveltimes are the same as for the direct arrival, (c) Shot and receiver at same side of diffraction point (back scatter), (d) CSP for $x_s = x_p$, i.e., $x_o = 0$ at $x_r = x_p$, (e) CRP for $x_r = x_p$, (f) CMP for $x_m = x_p$, for $|x_o| < 2x_p$ situation as in (c) occurs, (g) COP for $x_o = 0$ and $x_o = 2x_p$, (h) CTP in (x_s,x_r) for $t = 2t_p$, (i) CTP in (x_m,x_o) for $t = 2t_p$. t_p is traveltime from P to D. This event has the shape of a pyramid in (t,x_s,x_r). Compare with Fig. 3.12 and Fig. 3.14 to identify the parts that are in common with the direct arrival.

Using equation (3.4), equations (3.22a) and (3.22b) can be expressed in (x_m,x_o):

$$t(x_m,x_o) = |x_o|/V \qquad \text{for case 1,}$$

and

$$t(x_m,x_o) = |2x_m|/V \qquad \text{for case 2.}$$

(3.22c)

This peculiar event is described in various cross-sections in Figure 3.13d–i. Note that in the usual display of a shot record (CSP) zero offset ($x_s =$

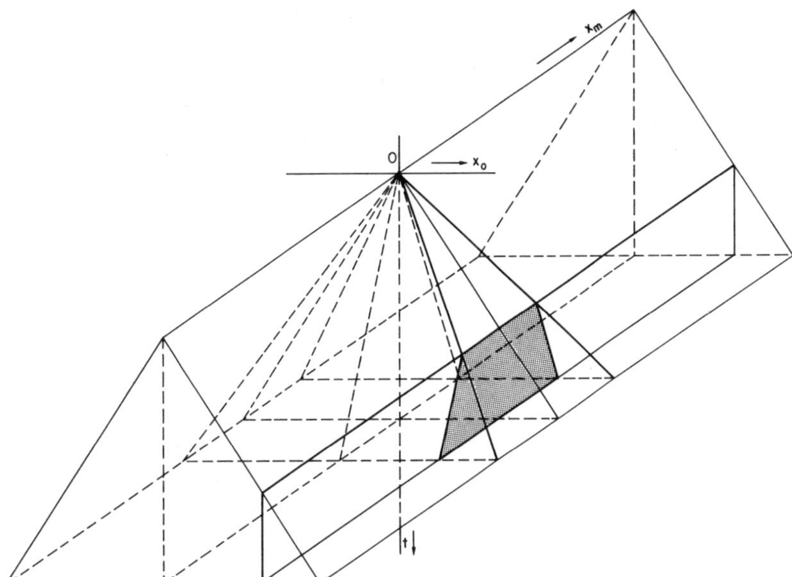

Fig. 3.14. Continuous wavefields in (t,x_m,x_o). Shown are the travel-time surfaces of the direct wave (tent shape) and of the diffraction from a diffractor at the seismic line at $x = 0$ (pyramid shape). A vertical cross-section through $x_o = C$ is indicated and can be compared with the COPs of Fig. 3.12e and Fig. 3.13g. The faces of the pyramid that do not coincide with the direct wave are the back scatter area for which shot and receiver are on the same side of the diffraction point.

x_r) would be at the center of the x_s axis for center-spread recording. Here the diffraction point is the center of the display. Combining the cross-sections of Figure 3.13 into a three-dimensional picture shows that the event has a pyramid shape in (t,x_m,x_o) and in (t,x_s,x_r). This pyramid shape is illustrated in Figure 3.14, which also shows that for case 1 two faces of the pyramid coincide with the direct wave, whereas the other two faces correspond to case 2. The traveltime surface of a diffraction point in (t,x_s,x_r) is called Cheops' pyramid in Claerbout (1985).

The edges of the pyramid correspond to the changeover points from case 1 to case 2, i.e., to shot or receiver located at the diffraction point, $x_s = 0$ or $x_r = 0$. Hence the diagonal vertical planes of the pyramid correspond to the CSP and CRP for shot and receiver at the origin, respectively. Figures 3.13d and e illustrate the case for shot and receiver located at $x_{s,r} = x_p$. The right-hand side of the traveltime curves corresponds to back scatter, the left-hand side to direct arrival.

Figure 3.15 shows a contour map of the traveltimes for a diffraction point at zero distance from the seismic line in a medium with constant velocity $V = 2000$ m/s. Once again, the pyramidal shape of the traveltime surface of this surface scatterer is illustrated.

The derivation of the apparent velocities in the various cross-sections is rather trivial. Interestingly the back scatter in the COP (Figure 3.13g)

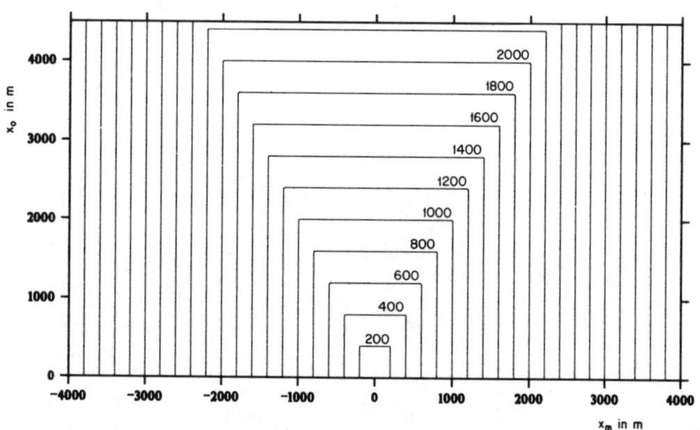

Fig. 3.15. Time contour map of a diffraction generated by a diffractor at the seismic line in a medium with constant velocity $V = 2000$ m/s.

has apparent velocity $\pm V/2$, which illustrates that the theoretical minimum apparent velocity as given in equation (3.14) can indeed occur.

3.8.3 A diffraction at some distance from the seismic line

Consider a diffraction point D at $(0,y,z)$ with $R = \sqrt{y^2 + z^2}$ in a medium with constant velocity V and a seismic line through $y = 0$, $z = 0$ (Figure 3.16a). If $t_0 = R/V$ is the traveltime from the origin to D, then the traveltime expression is

$$t(x_s,x_r) = \sqrt{t_0^2 + (x_s/V)^2} + \sqrt{t_0^2 + (x_r/V)^2}$$

$$= t_s + t_r. \tag{3.23}$$

Expression (3.23) describes a surface in (t,x_s,x_r). Note that $t(x_s,x_r) = t(x_r,x_s)$ in accordance with the principle of reciprocity. Note also that t_0 is the same for all points at the same distance R from the seismic line. The special case for $t_0 = 0$ was treated in the previous section.

The corresponding expression in (x_m,x_o) reads [using equation (3.4)]:

$$t(x_m,x_o) = \sqrt{t_0^2 + (x_m + x_o/2)^2/V^2} \tag{3.24}$$
$$+ \sqrt{t_0^2 + (x_m - x_o/2)^2/V^2}.$$

Various cross-sections through the traveltime surface are shown in Figures 3.16c–h. Figures 3.16c–f each correspond to one of the trajectories in (x_s,x_r) drawn in Figure 3.16b. Note that the curves in Figure 3.16c and Figure 3.16d are identical. Similar to Figures 3.13d and e, the zero-offset position in Figures 3.16c and d is at $x_r = x_p$, and $x_s = x_p$, respectively. For off-end shooting with $x_s > x_r$ the CSP would show only the curve left of $x_r = x_p$ in Figure 3.16c, whereas the CRP would show only the curve right of $x_s = x_p$ in Figure 3.16d.

The three-dimensional shape in (t,x_m,x_o) of the diffraction event can be plotted in a contour map. Figure 3.17 shows time contours for diffractors at varying distances from the seismic line. Only contours for x_o

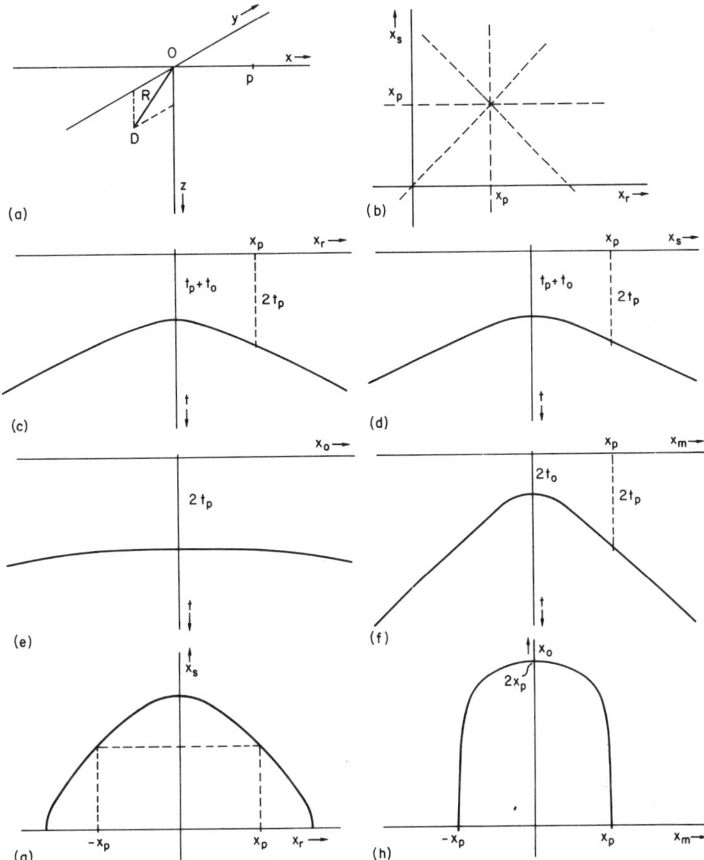

Fig. 3.16. Continuous wavefield of a diffractor in various cross-sections. (a) Geometry, D is diffractor at distance R from seismic line, P is some point on seismic line, (b) (x_s, x_r) plane with $x_s = x_p$, $x_r = x_p$, $x_m = x_p$ and $x_o = 0$, (c) CSP for $x_s = x_p$, (d) CRP for $x_r = x_p$, (e) CMP for $x_m = x_p$, (f) COP for $x_o = 0$, (g) CTP in (x_s, x_r) for $t = 2t_p$, (h) CTP in (x_m, x_o) for $t = 2t_p$. t_o is one-way traveltime 0 to D. t_p is one-way traveltime P to D.

> 0 are shown. The shape of the diffraction event evolves from a pyramid with sharp edges when the diffractor is at the surface on the seismic line (Figures 3.14 and 3.15) to a very smooth surface when the diffractor is at greater distance. For large distances from the midpoint to the diffractor the traveltime surface approaches asymptotically the faces of the pyramid shown in Figure 3.14.

An illustration of off-end shooting along a diffractor is given in Figure 3.18. The shooting geometry corresponding to Figure 3.18 is given as common shot trajectories in Figure 3.17a. The traveltimes in Figure 3.18 have been projected parallel to the x_o axis onto the $x_o = 0$ plane. The four (t, x_m) panels correspond to the four situations in Figure 3.17 with the restriction that $x_o < 3000$ m. Note the difference in shape of the traveltime curves between the right- and left-hand side of the pictures.

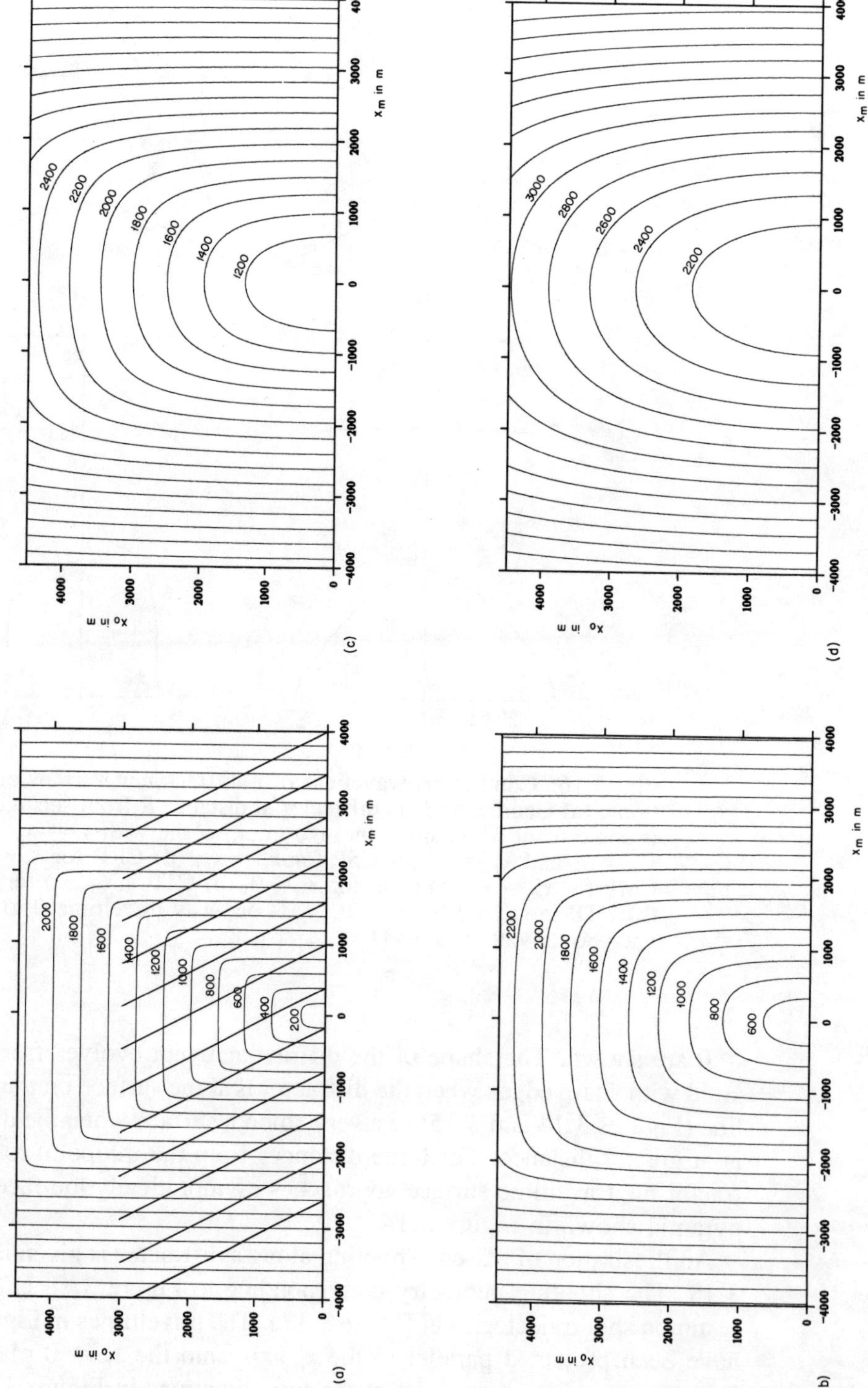

Fig. 3.17. Contour maps of traveltime surface of diffractions in a medium with constant velocity $V = 2000$ m/s. The distance from diffractor to seismic line is varied. (a) Distance = 100 m, this figure also shows the common shot trajectories corresponding to the shots in Fig. 3.18. (b) Distance = 500 m, (c) Distance = 1000 m, (d) Distance = 2000 m.

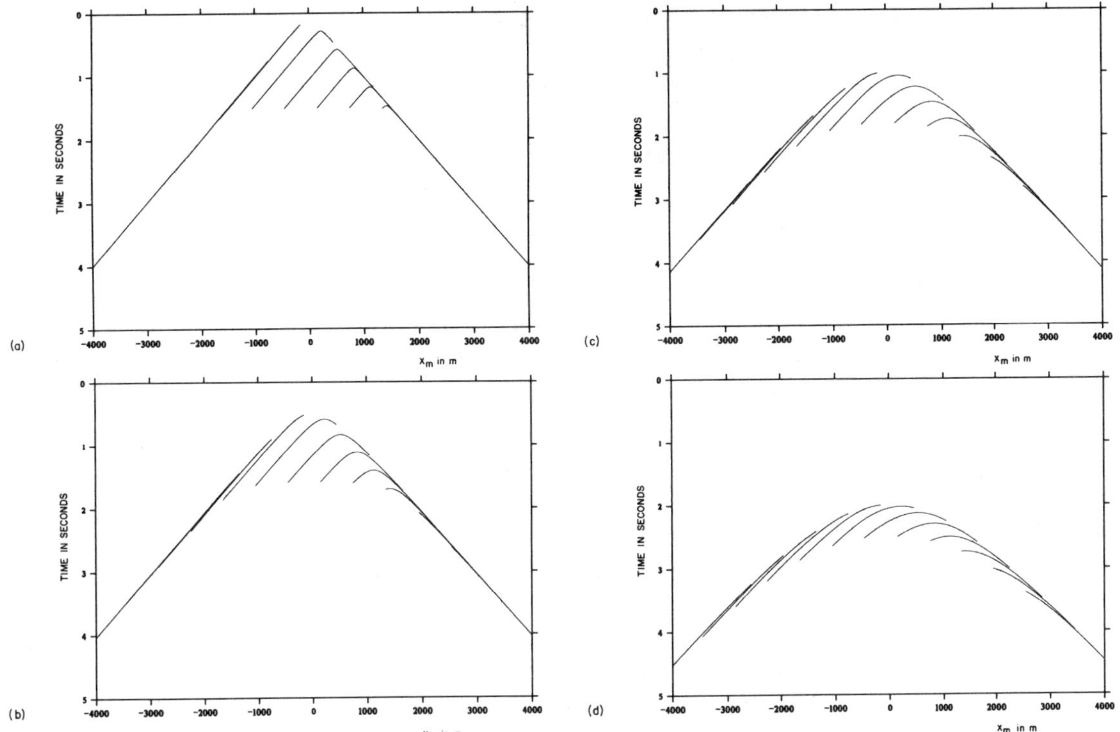

Fig. 3.18. Traveltimes of diffraction as recorded in an off-end shooting geometry with shots to the right of the spread ($x_s > x_r$), the common shot trajectories are shown in Fig. 3.17a. Maximum offset is 3000 m. Distance between shots is 600 m. Traveltimes are projected parallel to the x_o axis onto the $x_o = 0$ plane. Distance from diffractor to seismic line is varied. (a) Distance = 100 m, (b) Distance = 500 m, (c) Distance = 1000 m, (d) Distance = 2000 m.

The apparent velocities along the traveltime surface of a diffractor follow from equations (3.23) and (3.24) and are given by:

$$\frac{1}{V_s} = \frac{\partial t}{\partial x_s} = \frac{1}{V^2} \frac{x_s}{t_s}, \tag{3.25a}$$

$$\frac{1}{V_r} = \frac{\partial t}{\partial x_r} = \frac{1}{V^2} \frac{x_r}{t_r}, \tag{3.25b}$$

$$\frac{1}{V_m} = \frac{1}{V^2} \left(\frac{x_s}{t_s} + \frac{x_r}{t_r} \right), \tag{3.25c}$$

and

$$\frac{1}{V_o} = \frac{1}{2V^2} \left(\frac{x_s}{t_s} - \frac{x_r}{t_r} \right). \tag{3.25d}$$

Hampson (1987) points out that equations (3.25a and b) are a function of only one spatial variable, hence lines of constant V_s are parallel to the x_r axis and lines of constant V_r are parallel to the x_s axis.

Along the line $x_s = x_r$ in Figure 3.16b $t_s = t_r$. Hence it follows from equation (3.25c) that $1/V_m = 2/V_s = 2/V_r$, which illustrates that in the zero-offset panel the apparent velocities are half the corresponding velocities in CSP and CRP. Also, if shot and receiver both lie on the same side of the diffraction point with $|x_{s,r}| \gg R$, then it follows from equation (3.23) that t_s and t_r tend to $x_{s,r}/V$. Hence it follows from equation (3.25c) that V_m tends to $\pm V/2$ for $|x_m| \gg R$ [cf. case 2, equation (3.22b and d) in Section 3.8.2].

Equation (3.25d) can be used to demonstrate that diffractions have large apparent velocities in a CMP if shot and receiver both lie on the same side of the diffraction point. In the case of Figure 3.18 the whole spread lies on one side of the diffractor for $|x_m| > 3000$ m.

The properties of the traveltime surface of a single diffractor form the starting point for an analysis of the wavefield of a large number of random scatterers by various authors (Larner et al, 1983, Newman, 1984, and Lynn and Larner, 1989).

These authors discuss specifically the wavefield of scatterers that are randomly distributed over the sea floor. The diffractions produced by these scatterers travel mostly at a uniform propagation velocity, leading to traveltime surfaces for each diffractor as depicted in Figure 3.17, and described by equations (3.23) and (3.24).

Larner et al. (1983) model the response of a large number of random scatterers and show that stacking may enhance the response of some scatterers rather than suppress them. Newman (1984) computes the response of scatterers located at different azimuths with respect to a point on the seismic line and with different horizontal distances from that point. A suite of constant velocity stacks of the CMPs shows that there are always scatterers, usually in a narrow azimuth range, that stack almost perfectly. Using equation (3.24) one can easily verify that for $x_m = 0$ (azimuth is 90°) the diffractions stack perfectly for a stacking velocity equal to V, independent of distance from the seismic line. Hence, at shallow levels, the apexes of the diffractions may be reinforced by stacking. At deeper levels, farther away from the scatterer, the flanks of the traveltime surface may have an offset dependence similar to primary reflections and hence be reinforced by stacking. For these events shot and receiver are on the same side of the scatterer, so that V_m tends to $\pm V/2$. This explains why the scatterers are most conspicuous in the stack as two sets of parallel steeply dipping events with apparent velocity $\pm V/2$.

Lynn and Larner (1989) study the effect of cross-line shot patterns on the response of side-scattered energy. For the data investigated wide arrays had little or no effect. Lynn and Larner also present attenuation charts illustrating suppression by stacking as a function of stacking velocity, frequency, and azimuth. The charts show narrow azimuth bands of little suppression, confirming the results reported in Newman (1984).

3.8.4 Reflection of dipping horizon

Consider a dipping reflector with dip angle α overlain by a constant velocity medium.

The origin of the coordinate system is put at the intersection of the dipping plane and a dip-line at the surface (Figure 3.19a).

The traveltime expression follows immediately from the cosine rule:

$$t(x_s,x_r) = (x_s^2 + x_r^2 - 2x_r x_s \cos 2\alpha)^{1/2}/V, \qquad (3.26)$$

and substituting equation (3.4) leads to

$$t(x_m,x_o) = (4x_m^2 \sin^2\alpha + x_o^2 \cos^2\alpha)^{1/2}/V. \qquad (3.27a)$$

Figures 3.19c–h represent a number of cross-sections through the traveltime surface. Each graph in Figures 3.19c–f corresponds to one of the trajectories in Figure 3.19b. It follows from equation (3.27a) that the shape of the event in the common time panel in (x_m,x_o) is an ellipse.

Equation (3.27a) can also be written as

$$t(x_m,x_o) = \left[t_0^2 + \left(\frac{x_o \cos \alpha}{V} \right)^2 \right]^{1/2}, \qquad (3.27b)$$

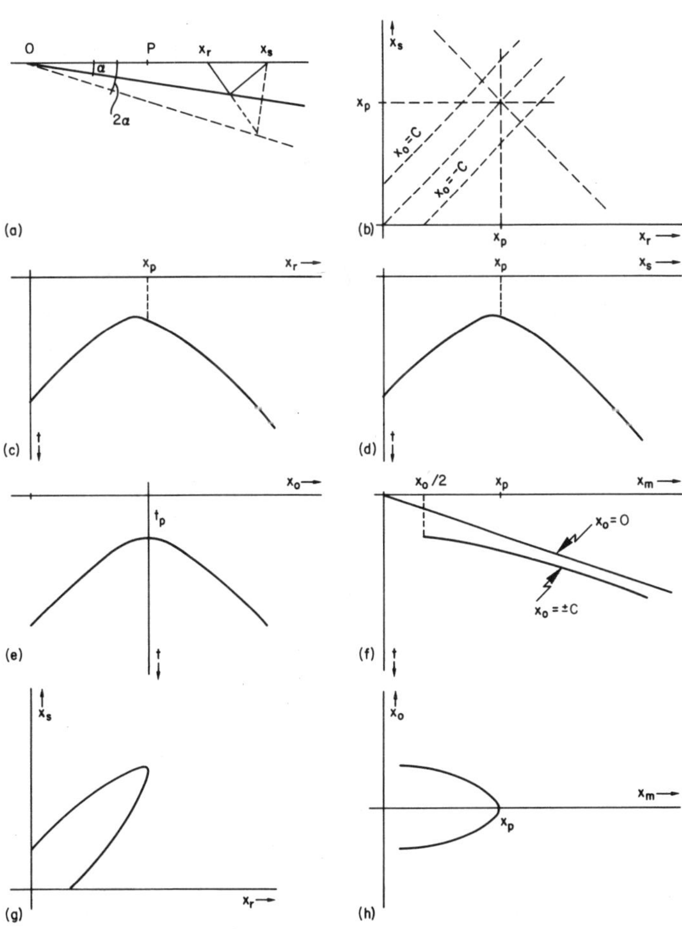

Fig. 3.19. Continuous wavefield of a dipping reflector in various cross-sections. (a) Geometry, P is some point on seismic line, (b) (x_s,x_r) plane with lines $x_s = x_p$, $x_r = x_p$, $x_m = x_p$ and $x_o = \pm C$, (c) CSP for $x_s = x_p$, (d) CRP for $x_r = x_p$, (e) CMP for $x_m = x_p$, (f) COP for $x_o = 0$ and $x_o = C$, (g) CTP in (x_s,x_r) for $t = t_p$, (h) CTP in (x_m,x_o) for $t = t_p$. No event can occur for $x_{s,r} < 0$, or for $x_m < 1/2\, x_o$.

with $t_0 = 2x_m \sin \alpha/V$ is the normal incidence time of the event. Equation (3.27b) is the well-known formula for a reflection hyperbola in the CMP.

For a COP with $x_o = C$ we can write equation (3.27a) as

$$t(x_m,x_o) = \left[\left(\frac{2x_m \sin \alpha}{V} \right)^2 + t_a^2 \right]^{1/2} , \qquad (3.27c)$$

with $t_a = C \cos \alpha/V$ being the apex of the hyperbola at $x_m = 0$. Note, however, that the hyperbola only exists for $x_m > 1/2\ C$. The hyperbola degrades to a straight line $t = 2x_m \sin \alpha/V$ for zero offset.

Figure 3.20 shows contour maps of events with varying dips in a constant velocity medium. An overlay of the trajectories of the CSP and

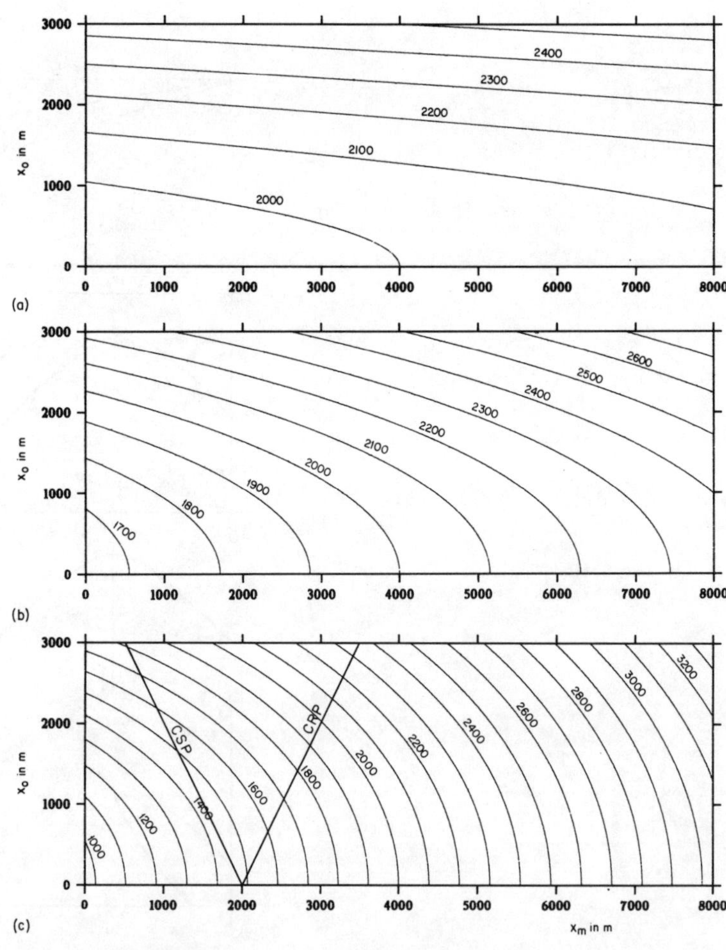

Fig. 3.20. Time contour maps of dipping events in a medium with constant velocity, only $x_o > 0$ is shown. $V = 2000$ m/s. (a) Dip = 1°, (b) Dip = 5°, (c) Dip = 15°. The trajectories of CSP and CRP at $x = 2000$ illustrate the difference in apparent dip seen by these panels for the case of off-end shooting with $x_s > x_r$.

CRP in Figure 3.20c illustrates the difference in apparent dip seen by these panels for the case of off-end shooting with $x_s > x_r$ (also: Section 4.6.9).

The apparent velocities in the four space time panels follow from equations (3.26) and (3.27a):

$$\frac{1}{V_r} = \frac{1}{V^2 t} (x_r - x_s \cos 2\alpha),$$

$$\frac{1}{V_s} = \frac{1}{V^2 t} (x_s - x_r \cos 2\alpha),$$

$$\frac{1}{V_m} = \frac{1}{V^2 t} 4x_m \sin^2 \alpha,$$

and

$$\frac{1}{V_o} = \frac{1}{V^2 t} x_o \cos^2 \alpha.$$

The behavior of dipping events in the CTP is further illustrated by model and real data examples in the next section.

3.9 Common Time Panels

3.9.1 Model data

Modeling programs can be used to analyze the behavior of more complicated models. Figure 3.21 shows a two-dimensional model consisting of one horizontal and four dipping reflectors. A ray-tracing program was used to construct the prestack data set corresponding to the model. Figure 3.22 shows some CMPs taken at regular intervals along the seismic line. Figure 3.23 shows two horizontal cross-sections through the data set in the (x_m, x_o) coordinate system. In Figure 3.23a event 2 does not change its offset position much, because it is not dipping steeply. In the top right-hand corner event 3 just appears. In Figure 3.23b, at 1800 ms,

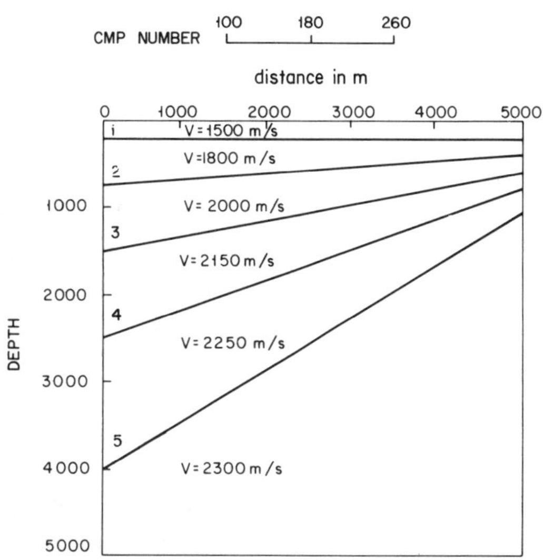

Fig. 3.21. Model of various dipping events. The CMP scale indicates the range of full coverage.

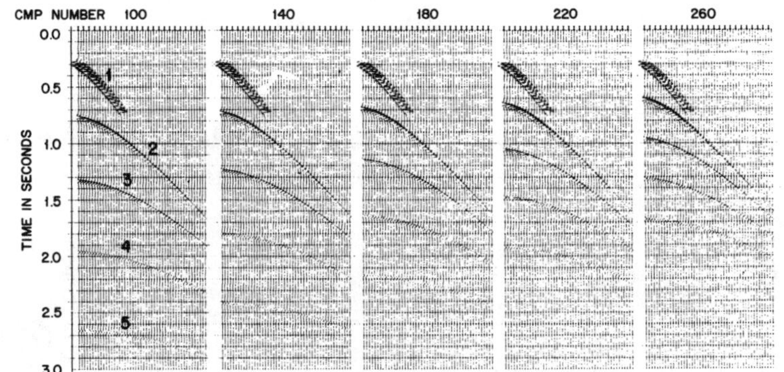

Fig. 3.22. Some common midpoint panels corresponding to the model of Fig. 3.21.

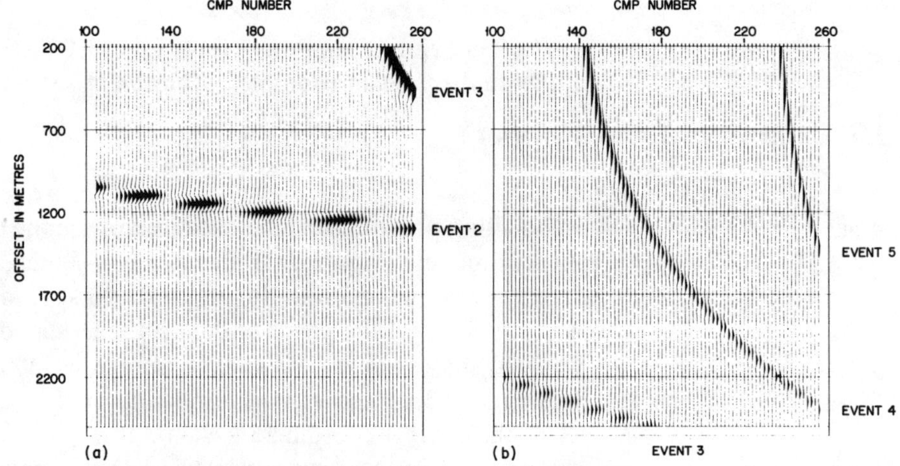

Fig. 3.23. Common time panels corresponding to the model of Fig. 3.21. CMP spacing is 25 m. Offset interval is 50 m. (a) CTP for $t = 1000$ ms, (b) CTP for $t = 1800$ ms. Note effect of undersampling in events 2 and 3.

event 3 is just visible in the bottom left-hand corner. Event 4 in Figure 3.23b illustrates the nearly elliptic behavior of a dipping event. Use of Figure 3.22 helps to understand the behavior of the various events in Figure 3.23.

As a preamble to the discussions on sampling in Chapter 4 we can deal here with the strange amplitude behavior of event 2 in the common time panel of Figure 3.23a. This amplitude varies strongly whereas it should be constant. The amplitude variation is caused by undersampling (sampling interval is 50 m) of the event as a function of x_o. In combination with the smoothing of the display system this undersampling leads to the typical football shapes in the figure. If the sampling interval is halved a dramatic improvement of the CTPs results (Figure 3.24a,b,c).

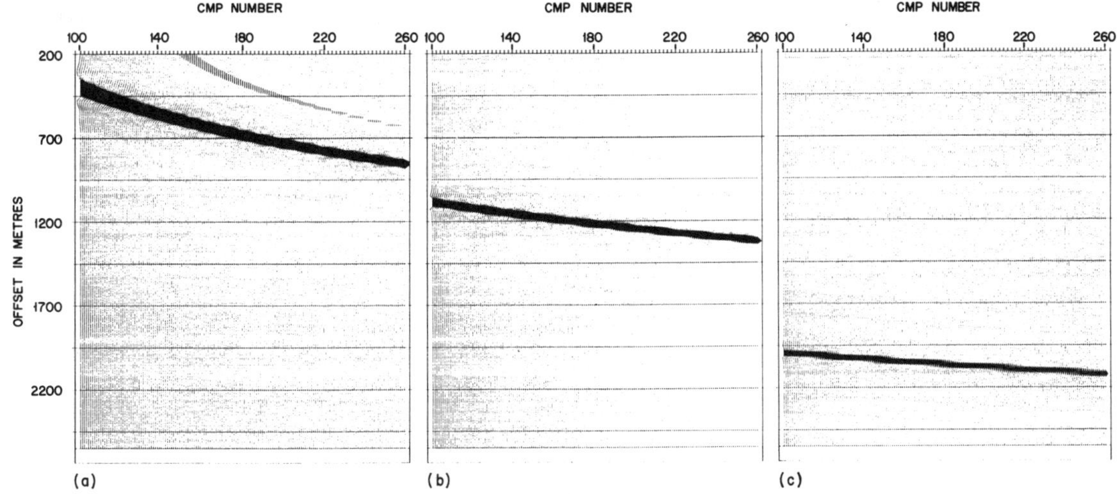

Fig. 3.24. Common time panels showing only the reflection of reflector 2 of model in Fig. 3.21. CMP spacing is 12.5 m. Offset interval is 25 m. (a) CTP for $t = 800$ ms, (b) CTP for $t = 1000$ ms, (c) CTP for $t = 1400$ ms.

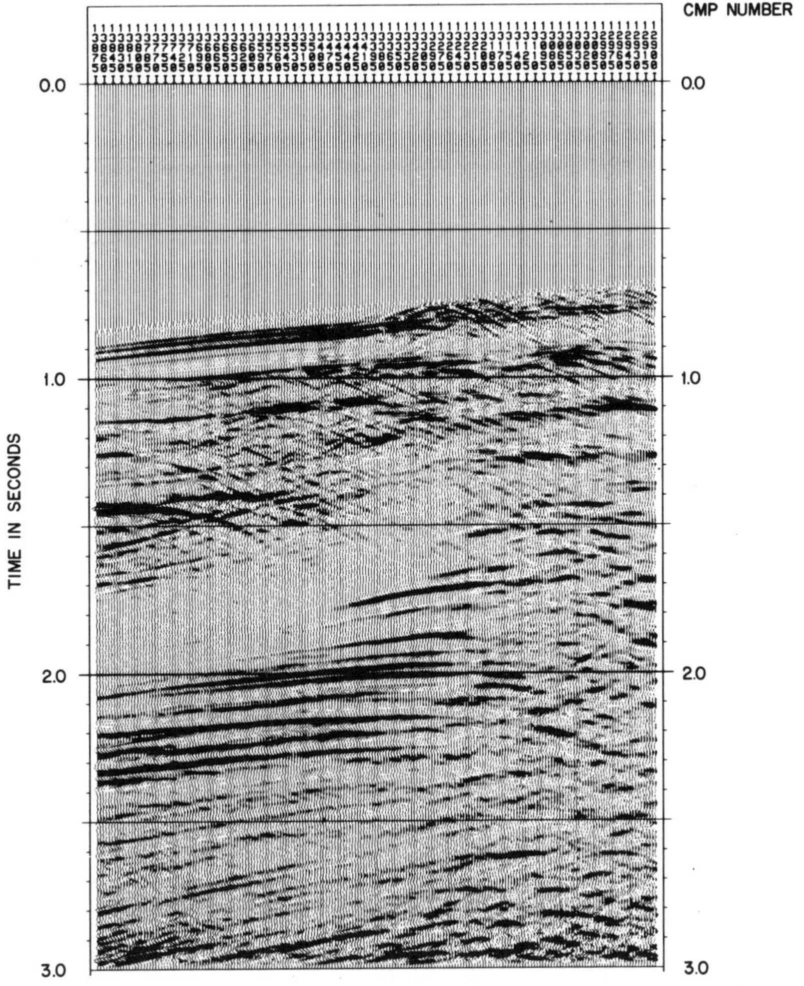

Fig. 3.25. 48-fold CMP stack showing some short high-amplitude events. The strong reflection between 1700 and 1800 ms is analyzed prestack in Figs. 3.26–3.28.

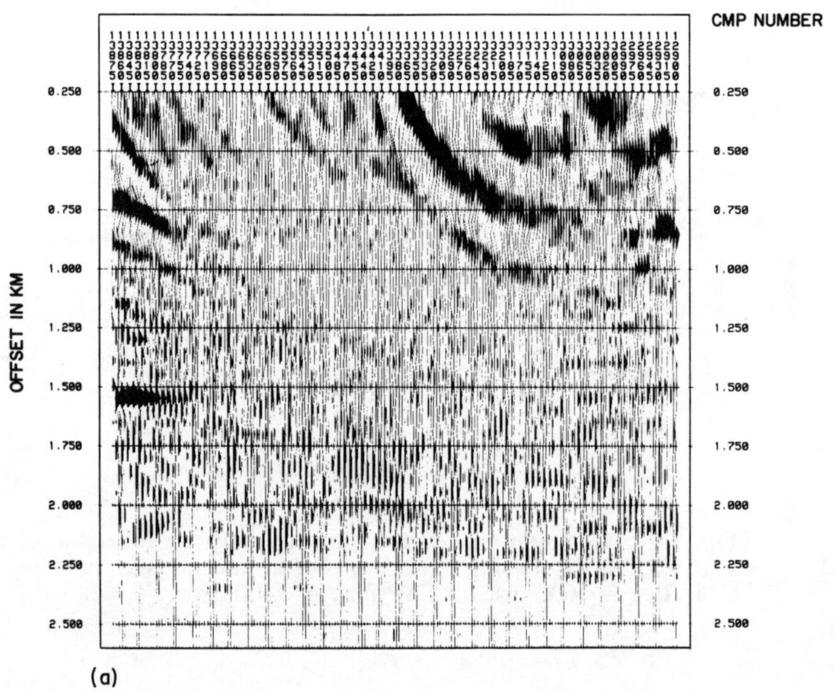

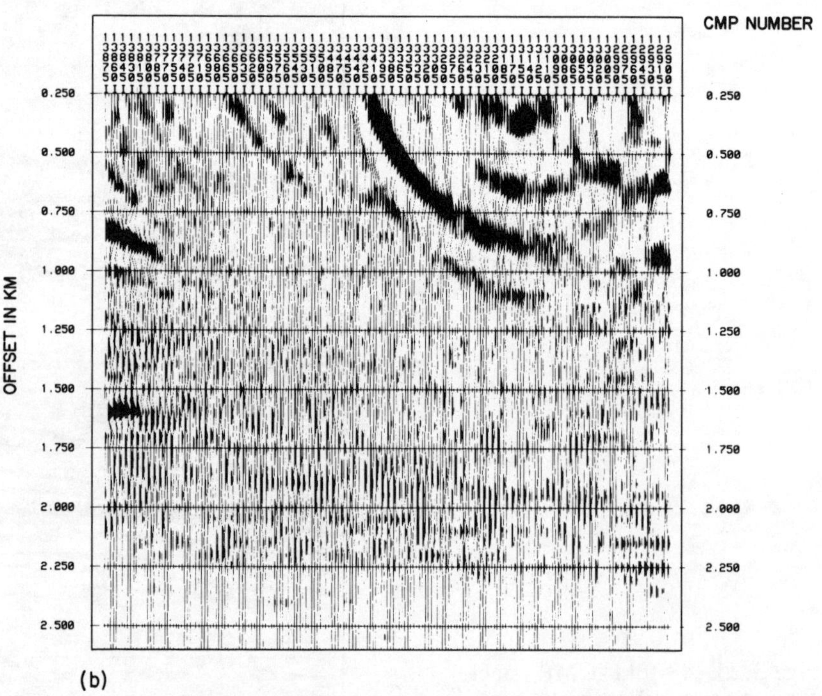

Fig. 3.26. Common time panels before NMO correction. (a) $t = 1750$ ms, (b) $t = 1770$ ms.

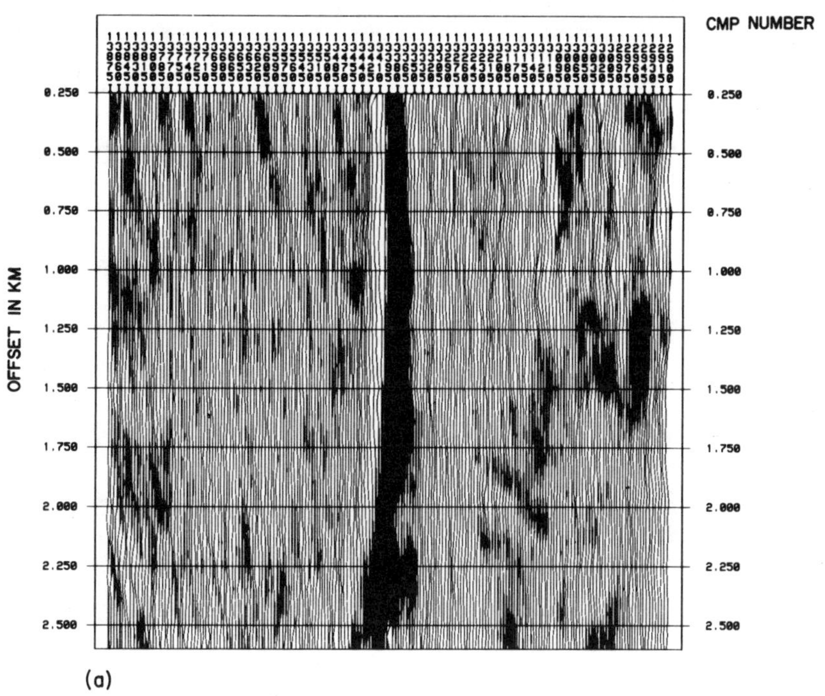

(a)

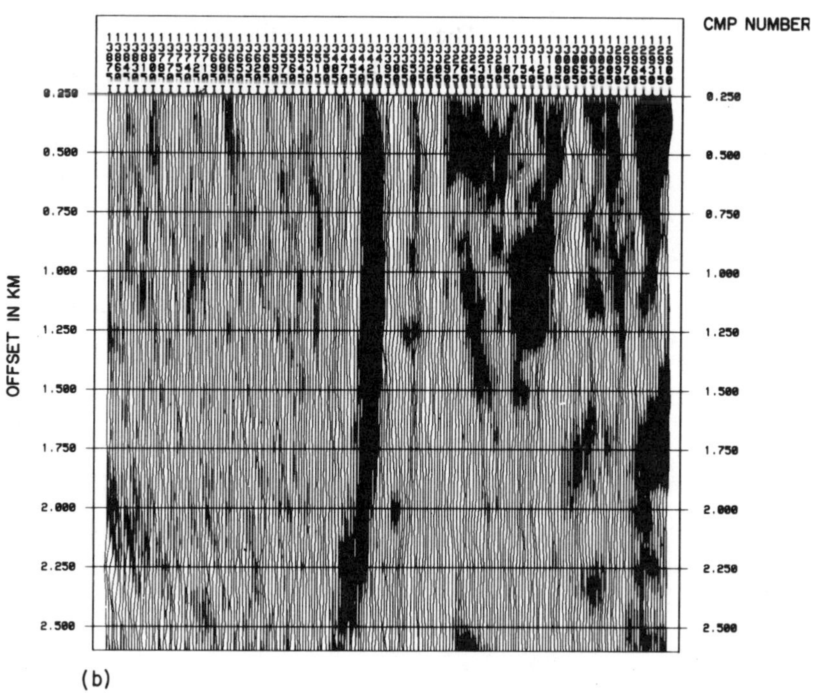

(b)

Fig. 3.27. Common time panels after NMO correction. (a) $t = 1750$ ms, (b) $t = 1770$ ms.

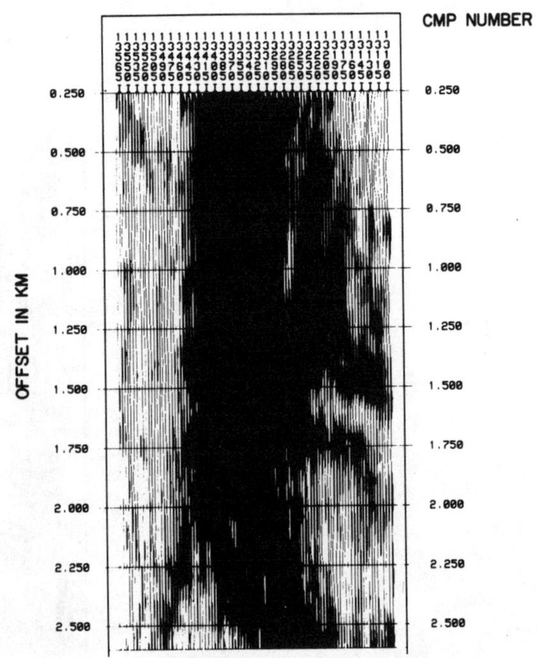

Fig. 3.28. Common horizon panel for reflection discussed in Fig. 3.25.

3.9.2 Real data

The seismic line the stack of which is shown in Figure 3.25 illustrates the use of the common time panel. Between 1700 and 1800 ms a dipping strong reflection stands out against a quiet background. Figure 3.26 shows CTPs of this line at 1750 ms and at 1770 ms. The dipping event stands out in these pictures in much the same way as event 4 in Figure 3.23b. Figure 3.27 shows the corresponding CTPs after NMO correction. Note that the strong event is now curved to the left at larger offsets. This orientation means that the event was somewhat overcorrected (too slow a stacking velocity). With larger stacking velocities the event could have been curved to the right at larger offsets, as in Figure 3.26. These results illustrate the diagnostic value of common time panels of NMO-corrected data.

If both moveout and structure are flattened, a common horizon panel results, as shown in Figure 3.28. This representation is useful for the investigation of offset-dependency of seismic amplitudes.

Chapter 4
THE RECORDED WAVEFIELD

4.1 Recording Effects and Earth-Response Effects

Section 3.1 defined the hypothetical continuous wavefield. Chapter 4 deals with the recording (sampling along the surface and in time) of this continuous wavefield, defined as the *recorded wavefield*.

The recording introduces a host of effects most of which can be described by a convolution in one or two of the independent variables. Section 4.1.1 lists the more important effects.

The recording effects must be distinguished from unwanted parts of the earth response. Section 4.1.2 lists some unwanted earth-response effects.

Some recording effects can be removed using their surface consistency. Surface consistency is discussed in Section 4.1.3.

4.1.1 Recording effects on the continuous wavefield

The recorded wavefield $u(t,x_s,x_r)$ can be described as a convolution of the original continuous wavefield with a number of effects caused by the recording action:

$$u(t,x_s,x_r) = W(t,x_s,x_r) * R(t,x_r) * S(t,x_s) * S_\Delta(t) * S_\Delta(x_s)$$
$$* S_\Delta(x_r) * A(t) * I(t) * B[x_o] * B[x_m] * p(x_s)$$
$$* p(x_r) + n(t,x_s,x_r) \tag{4.1}$$

in which

$$
\begin{aligned}
W(t,x_s,x_r) &= \text{continuous wavefield,} \\
R(t,x_r) &= \text{receiver response, including geophone-ground coupling,} \\
S(t,x_s) &= \text{source wavelet and strength,} \\
S_\Delta &= \text{sampling operator,} \\
A(t) &= \text{antialias filter,} \\
I(t) &= \text{instrument response,} \\
B[x_o] &= \text{spread length box function,} \\
B[x_m] &= \text{line length box function,} \\
p(x_s) &= \text{shot pattern,} \\
p(x_r) &= \text{receiver pattern, and} \\
n(t,x_s,x_r) &= \text{additive ambient noise.}
\end{aligned}
$$

This description ignores nonadditive noise. Other effects that are not included above are: variations in shot and receiver depths, nonelastic effects at the source, missing shots and receivers, directivity of the source wavelet, and probably more.

4.1.2 Unwanted parts of the earth response

In conventional P-wave recording the aim is to create a zero-offset section consisting of primary compressional reflections only. Unwanted parts of the earth response are in that case: source and receiver ghosts, near-surface effects, e.g. variations in sea-bottom properties, shot-generated noise (see remark in Section 3.4), multiples and reverberations, and converted waves.

Seismic data may also be gathered with other objectives in mind, and some of these "unwanted" effects may provide useful information: near-surface effects could give information useful in site surveys. Converted waves are exploited in PSV recordings.

4.1.3 Surface consistency

"Surface consistency", a term commonly used in connection with statics, means that the time delay for a trace at (x_s,x_r) can be written as

$$t(x_s,x_r) = t_s(x_s) + t_r(x_r) \qquad (4.2)$$

in which $t_s(x)$ and $t_r(x)$ represent shot and receiver statics, respectively.

Note that surface consistency of statics does not imply reciprocity. If statics were reciprocal, it would mean

$$t_s(x_s) + t_r(x_r) = t_s(x_r) + t_r(x_s),$$

which can be valid for all values of $x_{s,r}$ only if

$$t_s(x) = t_r(x) + C.$$

The main characteristic of equation (4.2) is the *separability* of the statics into shot effect and receiver effect. Thus surface consistency can be generalized in the following definition: *Surface consistency is the property that recording effects on the wavefield can be separated into shot effects and receiver effects.*

For instance, the source wavelet $S(t,x_s)$, [see equation (4.1)], is present in all traces with the same x_s. Thus we can talk about surface-consistent wavelet extraction and deconvolution. Equalization algorithms (Section 5.3) should also exploit the property of surface consistency.

4.2 Spatial Sampling

The reason the continuous wavefield was termed "hypothetical" in Section 4.1 is because an infinite number of sources would be needed to generate the field and an infinite number of receivers would be needed to record the field. In other words, spatial sampling in x_s and x_r is (as yet) a physical necessity. Shot and receiver intervals should be selected such that the reconstruction of the continuous wavefield from the recorded samples is possible. If it is possible to reconstruct the continuous wavefield from the recorded samples, then the data are said to be *properly sampled.*

In Section 3.7 we saw that for each spatial coordinate x_i, there exists a maximum wavenumber $|k_i|_{\max}$, provided (1) $W_{tsr}(f,k_s,k_r) = 0$ for $f > f_{\max}$, and (2) there is a $V_{\min} > 0$ [equations (3.20) and Figure 3.10]. According to the Nyquist condition (Nyquist, 1928) or the Shannon sampling theorem (Shannon, 1949; Jerri, 1977) proper sampling (or alias-free sampling) of each coordinate of the wavefield is achieved if

$$\Delta t \leq \frac{1}{2f_{max}},$$

$$\Delta x_s \leq \frac{1}{2|k_s|_{max}} = \frac{V_{min}}{2f_{max}},$$

$$\Delta x_r \leq \frac{1}{2|k_r|_{max}} = \frac{V_{min}}{2f_{max}},$$

$$\Delta x_o \leq \frac{1}{2|k_o|_{max}} = \frac{V_{min}}{2f_{max}},$$

and

$$\Delta x_m \leq \frac{1}{2|k_m|_{max}} = \frac{V_{min}}{4f_{max}}. \tag{4.3}$$

We call Δx_i [using the equal sign in inequalities (4.3)] the *basic sampling interval* for x_i. The corresponding maximum frequency and wavenumbers are called Nyquist frequency f_N and Nyquist wavenumber k_{i_N}, respectively, i.e., $f_N = 1/2\Delta t$ and $k_{i_N} = 1/2\Delta x_i$.

The reciprocity theorem has led us by way of equation (3.14) to the result that the basic sampling intervals for shots and receivers are the same. The basic offset sampling interval Δx_o is equal to Δx_s and Δx_r, whereas the basic midpoint sampling interval Δx_m is half the other basic sampling intervals.

$|k_{s,r}|_{max}$ may be smaller than would follow from $|k_{s,r}|_{max} = f_{max}/V_{min}$ if events with the smallest apparent velocities do not carry frequencies up to f_{max}. Occurrence of this situation can be analyzed using microspreads (Section 4.10.1). In the following discussion we assume that $|k_{s,r}|_{max}$ is proportional to f_{max}.

The frequency spectrum generated by the source is usually under limited control, with Vibroseis as a notable exception. Dynamite may generate very high frequencies. The frequency spectrum of ambient noise may also contain energy at high frequencies. Therefore, to ensure that f_{max} of the recorded wavefield is not larger than f_N an analog antialias filter $A(t)$ must be applied prior to sampling. The antialias filter suppresses the energy for frequencies larger than f_N.

According to equation (4.3) all basic sampling intervals are inversely proportional to f_{max}, so that the antialias filter in t effectively acts as an antialias filter in x_s and x_r as well. This is quite fortunate, because application of proper antialias filters $A(x_s)$ or $A(x_r)$ directly is practically impossible.

If the continuous wavefield is sampled according to equation (4.3) then the Nyquist frequency and wavenumbers are equal to the maximum recorded frequency and wavenumbers, so that aliasing, temporal or spatial, does not occur.

If we want to record up to 75 Hz in a medium with $V_{min} = 1500$ m/s, then equation (4.3) leads to

$$\Delta t = 6\ 2/3 \text{ ms},$$

$$\Delta x_s = \Delta x_r = 10 \text{ m (and } \Delta x_o = 10 \text{ m, } \Delta x_m = 5 \text{ m)}.$$

Usually, and certainly in land data acquisition, $V_{\min}$ is much smaller, perhaps 150 m/s; in that case

$$\Delta t = 6\ 2/3 \text{ ms},$$

$$\Delta x_s = \Delta x_r = 1 \text{ m (and } \Delta x_o = 1 \text{ m, } \Delta x_m = 0.5 \text{ m)}.$$

Selecting spatial sampling intervals as small as 10 m, let alone 1 m, is as yet uncommon which means that a certain amount of spatial aliasing is virtually always present. Field patterns are an attempt to reduce the amount of aliasing.

Introducing the basic *signal* sampling interval in addition to the basic sampling interval is useful. The basic signal sampling interval is the sampling interval that would properly sample the desired signal energy (usually compressional-wave energy, including diffractions), but not the groundroll.

Obviously the continuous wavefield can be sampled properly, at least theoretically. However, usually there are disturbances due to sampling irregularities and ambient noise effects. These disturbances are discussed further in the next section.

4.3 Disturbances of the Recorded Wavefield

4.3.1 Sampling irregularities

Temporal sampling of the continuous wavefield can be performed with great accuracy, whereas spatial sampling is liable to errors that reduce the dynamic range of the seismic system significantly. Newman and Mahoney (1973) list element effectiveness, element positioning, ground coupling, and local heterogeneities as sampling irregularities. I define *sampling noise* as the net effect of all these irregularities on the recorded wavefield. Ongkiehong and Huizer (1987) reported an rms spread in effective sensitivity of well-planted geophones of 17.6 percent and 26 percent for two investigated cases. Variations in hydrophone sensitivity will also lead to sampling noise.

At the shooting end variations in shot strength (hole conditions in dynamite shooting, surface conditions for Vibroseis recording) will also create sampling noise. The effect of these disturbances is that the wavenumber spectra are no longer bounded to the triangular areas as indicated in Figure 3.8, but are spread over all wavenumbers.

Variations in shot strength are surface consistent with respect to the shot location. Assuming that the response of the geophone once planted does not vary, the effect of irregular geophone responses is also surface consistent, whereas for towed streamers variation in hydrophone sensitivity is *channel consistent*.

A reduction of the wavefield distortion may be found in redundant sampling, e.g. planting more geophones close together.

4.3.2 Ambient noise

Ambient noise may be defined as the energy recorded at the receivers but generated by an external source. The time frame of this noise is independent from that of the earth response. If the experiment is repeated, the same earth response will be recorded (provided the earth has not changed, see remark on repeatability in Section 3.1), but different ambient noise. Reciprocity does not apply to ambient noise.

Ambient noise can be divided into (1) noise that is part of the elastic wavefield as recorded and therefore subject to the same elastic properties of the subsurface as the earth response, and into (2) noise that is not part of the elastic wavefield. Examples of the first type are traffic noise, engine noise (except that part of traffic and engine noise that travels through the air), cattle and men walking along the spread, etc.; examples of the second type are instrument noise, noise recorded by geophones exposed to wind energy, cattle nibbling at the geophones, energy of water turbulences (depth controllers), etc.

Ambient noise of the first type has some peculiar properties. Within a common shot panel the ambient noise will have the same properties as the continuous wavefield, notably the minimum apparent velocity $|V_r|_{min}$ will apply. Within a CRP, however, the ambient noise is unrelated and energy exists in the whole (f,k_s) panel. Figure 4.1 illustrates this and shows that antialias filtering in t (creating an f_{max}) does not generate a $|k_s|_{max}$ (cf. Figure 3.8b). The k_s wavenumber spectrum extends from $-\infty$ to $+\infty$. The same reasoning applies to (f,k_m) and (f,k_o). If the earth response is sampled at least at the basic signal sampling interval, part of the ambient noise can be removed in processing by application of a velocity filter (rejecting all velocities below the minimum apparent velocity of the desired signal) in (f,k_s), (f,k_m) and/or (f,k_o) without hurting the desired earth response.

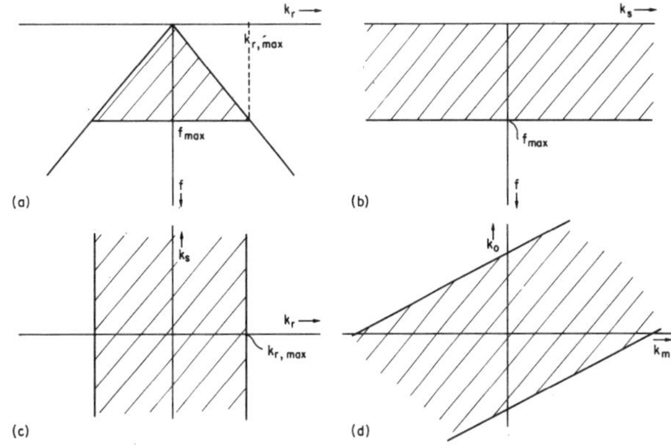

Fig. 4.1. Properties of ambient noise propagating through the elastic medium. (a) Energy is bounded to a triangle in (f,k_r), (b) In (f,k_s) panel no $|k_s|_{max}$ exist, (c) Common frequency panel in (k_s,k_r) for $f = f_{max}$, (d) Common frequency panel in (k_m,k_o) for $f = f_{max}$.

Ambient noise of the second type has the same properties as noise of the first type, but additionally is incoherent from trace to trace in the CSP. Hence, this noise can be tackled by velocity filtering in all four (f,k_i) panels.

4.4 Description of Shooting Geometry

Figure 4.2 is meant to promote a better understanding of the relation between a particular shooting configuration and the two coordinate systems (t,x_s,x_r) and (t,x_m,x_o). The example is an off-end shooting geometry. In the upper part of Figure 4.2 each dot represents a seismic trace with shot at $x = x_s$ and receiver station at $x = x_r$. In the middle part the actual shot and receiver positions are indicated along the x coordinate of the seismic line. For simplicity only eleven shots with eight receivers each have been drawn. The shot move-up dx_s is equal to the station spacing dx_r; the distance from each shot to the nearest receiver station equals twice this spacing. In the lower part of the figure the shooting configuration is shown in the (x_m,x_o) coordinate system. The upper part of Figure 4.2 is sometimes referred to as *surface diagram,* and the lower part as *subsurface diagram* (Taner et al, 1974).

The (x_s,x_r) coordinate system (surface diagram) is not very convenient

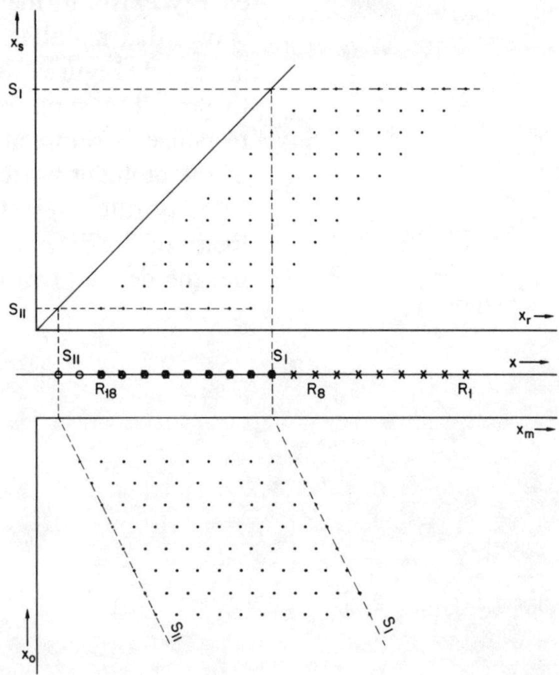

Fig. 4.2. Regular off-end shooting geometry, shots moving from right to left. Shot S_1 recorded by receivers R_8 . . . R_1, etc. Top: (x_s,x_r) coordinate system, each dot represents a trace in this system, Middle: Shots and receiver positions along seismic line, Bottom: (x_m,x_o) coordinate system, each dot represents a trace in this system. Each trace in (x_s,x_r) corresponds to a trace in (x_m,x_o). According to equation (3.3) all traces have negative offset x_o.

for describing a shooting configuration. In (x_m,x_o) the description is easier owing to the spread length acting as a window in x_o. In headboards of land seismic sections the subsurface diagram is often used.

The lower part of Figure 4.2 also illustrates the well-known odd/even effect. Each CMP contains only four different offsets, whereas there are eight receivers for each shot. Hence adjacent CMPs have different offset configurations.

The distance between neighboring traces in the CMP is twice as large as the distance between receivers. Also, in the COP the distance between neighboring traces is twice as large as the distance between adjacent CMPs. For proper sampling in (x_s,x_r) this leads to undersampling in (x_m,x_o). This sampling paradox is discussed further in the next section.

4.5 The Sampling Paradox

Now consider a properly sampled data set i.e., a data set for which the shot sampling interval dx_s and the receiver sampling interval dx_r both equal the basic sampling interval Δx. In Figure 4.3a points in (x_s,x_r) are plotted at distances equal to the basic sampling interval $\Delta x_s = \Delta x_r = \Delta x$. As in Figure 4.2 each dot at (x_s,x_r) represents a trace with the source at $x = x_s$ and the receiver station at $x = x_r$. Consider two adjacent points P and Q along a common offset trajectory. The trace at P is recorded by shot 1 with $x_s = 0$, the trace at Q is recorded by shot 2 with $x_s = \Delta x$.

In (x_s,x_r) we have P:$(0,\Delta x)$ and   Q:$(\Delta x,2\Delta x)$, while

in (x_m,x_o) we have P:$(\Delta x/2,-\Delta x)$ and   Q:$(3\Delta x/2,-\Delta x)$,

which shows that adjacent points in the COP lie at a distance $3\Delta x/2 - \Delta x/2 = \Delta x$ that is twice as large as the basic sampling interval Δx_m in x_m. The same applies to CMPs: the interval between traces with the same x_m is twice as large as the basic offset sampling interval Δx_o. Thus there seems to exist a *sampling paradox:* while the continuous wavefield is

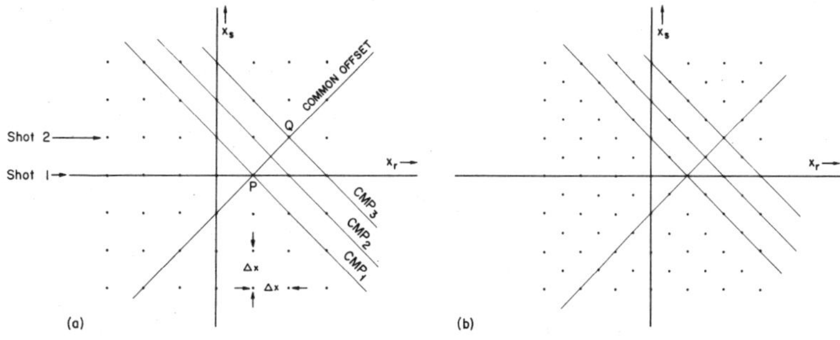

Fig. 4.3. Illustration of sampling paradox; each dot represents a trace in (x_s,x_r). (a) Proper sampling in (x_s,x_r) leads to undersampling in (x_m,x_o). CMP_2 is not represented in the offset panel defined by P and Q. (b) Proper sampling in (x_m,x_o) requires twice as many shots with receiver positions in between the original ones.

properly sampled in (x_s,x_r), i.e., at any point in (x_s,x_r) the value of the wavefield can be reconstructed from the sampled points, it is undersampled in (x_m,x_o). We would need twice as many traces to achieve proper sampling in (x_m,x_o). Proper sampling in (x_m,x_o) could be achieved by halving the shot interval (Figure 4.3b). Then alternate shots would be recorded at alternate receiver positions.

The sampling paradox can be solved by realizing that we used the one-dimensional Shannon sampling theorem in Section 4.2 to derive a sampling interval for each individual coordinate. However, the generalized N-dimensional sampling theorem, as formulated in Petersen and Middleton (1962) states that the most efficient sampling lattice (i.e., requiring minimum sampling points per hypervolume) is not generally rectangular.

If the two-dimensional spectrum is zero outside a square as in Figure 3.9a, then square sampling as depicted for (x_s,x_r) at the top of Figure 4.2 is most efficient. However, if the two-dimensional spectrum is zero outside a parallelogram around the origin as in Figure 3.9b, then oblique sampling as depicted for (x_m,x_o) in the bottom part of Figure 4.2 is most efficient. Hence the most efficient way to sample properly (x_m,x_o) happens to be square sampling of shots and receivers. Reconstruction of any point in (x_m,x_o) is then possible using a two-dimensional interpolation procedure. Thus the points added to Figure 4.3a to get Figure 4.3b do not need to be recorded because they can be computed.

Even if the total wavefield is properly sampled in (t,x_s,x_r) and therefore also in (t,x_m,x_o), the fact remains that any individual COP or CMP is undersampled. As a consequence various processes in the COP or CMP, e.g., computation of the (τ,p_m) or (τ,p_o) transform, would suffer from aliasing problems. Therefore the extra points (traces) in Figure 4.3b with respect to Figure 4.3a should indeed be determined by reconstruction or actual recording in case such processes in COP or CMP are to be carried out without the risk of aliasing effects. The way in which the missing traces can be reconstructed depends on the selected acquisition technique.

In off-end shooting with $dx_s = dx_r = \Delta x$ the reconstruction or "dealiasing" can be carried out only by interpolation in the processing center. This procedure is discussed in Section 5.6. An alternative is to oversample shots by choosing $dx_s = 1/2\ dx_r$.

In center-spread recording we can make use of the reciprocity principle [equations (3.9) and (3.10)] to obtain properly sampled (t,x_m) and (t,x_o) panels. In that case shots should be positioned half-way between receivers. A subset of this configuration is shown in Figure 4.4a. Each horizontal row of dots in Figure 4.4a represents a recorded shot. The row corresponds to a vertical column of circles representing the *reciprocity traces* [traces mirrored around $x_o = 0$, $W(t,x_m,x_o) = W(t,x_m,-x_o)$]. Together recorded traces and reciprocity traces form a properly sampled data set in (x_m,x_o). Knapp (1985) and Hobson (1985) also discuss this optimum center-spread shooting configuration.

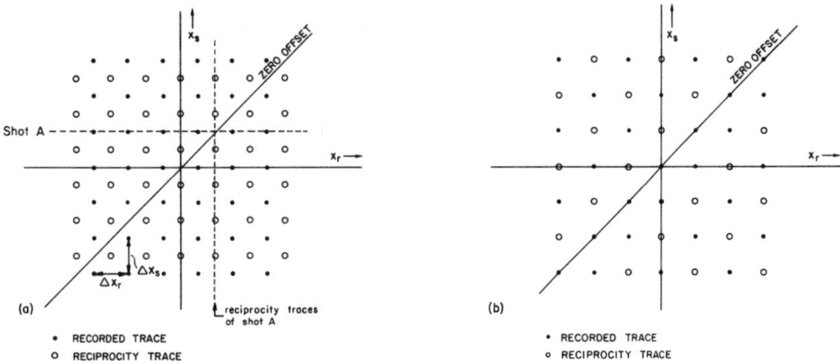

Fig. 4.4. Use of reciprocity principle in center-spread recording. Each recorded trace can be mirrored against $x_o = 0$ to form a reciprocity trace, (a) $dx_s = dx_r$ in center-spread recording leads to proper sampling in (x_m,x_o) (cf. Fig. 4.3b), provided shots are half-way between receiver positions, (b) Recording every other receiver can still lead to proper sampling in (x_s,x_r) (cf. Fig. 4.3a) making use of reciprocity, provided an asymmetrical (with respect to zero-offset) configuration is used with shots at receiver positions. The points shown in these figures are a subset of the actual shooting geometry, i.e., window effects as in Fig. 4.2 are not shown.

Proper sampling in (x_s,x_r) can also be achieved by making use of the reciprocity principle and shooting center spread in an asymmetrical configuration (Figure 4.4b). Rectangular sampling in (x_m,x_o) can then be achieved only in processing by interpolation. This solution is more theoretical than practical, though in a situation with sufficient receiver stations but insufficient recording channels it might be used.

This section discusses the sampling paradox for the theoretically ideal shooting configuration in which the basic sampling intervals are used. Note, however, that also in conventional data acquisition with dx_s equal to dx_r but much larger than the basic sampling interval Δx, CMP and COP are undersampled as compared to CRP and CSP (see also the lower part of Figure 4.2). If aliasing is restricted by the application of patterns (Section 4.6), then the discussion in this section applies also to conventional recording. In fact, the larger the spatial sampling intervals, the more serious is the problem of undersampling in COP and CMP.

4.6 Patterns

4.6.1 Need for patterns as antialias filters and as resampling operators

The term "pattern" is used throughout to indicate an arrangement of identical elements (sources or receivers) that form one source station or one receiver station. All sources in a *pattern* are fired simultaneously, and the signal detected by the receivers of a pattern is summed to record one trace. [Vibroseis source patterns are usually recorded in parts, and then summed (vertically stacked) in the field.] The elements of a one-dimensional pattern are usually equidistant, but other arrangements may be used. Unless otherwise stated, I discuss one-dimensional patterns layed out along the x coordinate (the seismic line). In the literature "array" is often used (Sheriff, 1984) rather than pattern. "Array" is used here only for groupings of air guns designed to create an optimum peak-to-bubble ratio using interaction between air guns of different sizes. Several such air gun arrays may be combined to form a marine air gun pattern.

In the past patterns were used as a means to suppress the unwanted part of the wavefield, especially groundroll. Microspreads (also called noisespreads), shot with small distances between receiver positions and without patterns, were used to measure the wavelengths with the greatest energy of the noise. The pattern lengths were designed to create full suppression for those wavelengths. The result was very long patterns that at the same time affected the desired part of the wavefield. The stacking process (Mayne, 1962) was introduced in the early 1950s by Mayne as the alternative to long patterns. Since then the stacking multiplicities have increased drastically while using ever-shorter receiver and shot patterns. The ultimate of this, shooting and recording at the basic sampling interval (defined in Section 4.2), is not yet common practice.

As long as seismic data acquisition is not carried out using the basic sampling interval, patterns are required. The patterns are not aimed so much to suppress noise, but more to act as a *spatial antialias filter* and as a *resampling operator*. Noise as well as signal are still suppressed by the short patterns, but the aim has become to prevent aliasing of noise *and* signal.

The action of a receiver pattern can be described by two steps (for each time sample):

(1) *Sample* the continuous wavefield at the position of the pattern elements. (This action is subject to the same sampling noise as described in Section 4.3.1.)

(2) *Add* the samples of one pattern (weighted or not) to form one output sample per pattern.

Conceptually the second step can be split into two parts:

(2a) *Convolve* a digital antialias filter with the sampled wavefield. The filter coefficients are equal for patterns that are not weighted.

(2b) Keep one output point of the convolution product for every receiver station and throw away the rest, i.e., *resample*.

The distance between sampling points after resampling, the receiver distance dx_r, determines the new Nyquist wavenumber $k_{r_N} = 1/2dx_r$. The action of shot patterns can be described in a similar way.

The Nyquist wavenumber determines the desired response of the antialias filter of step 2a. The desired zero-phase impulse response of an antialias filter is shown in Figure 4.5a. The cut-off wavenumber k_N is at $k_{s,r} = 1/2d$. The filter has a passband for $k_{s,r} < 1/2d$ and a reject

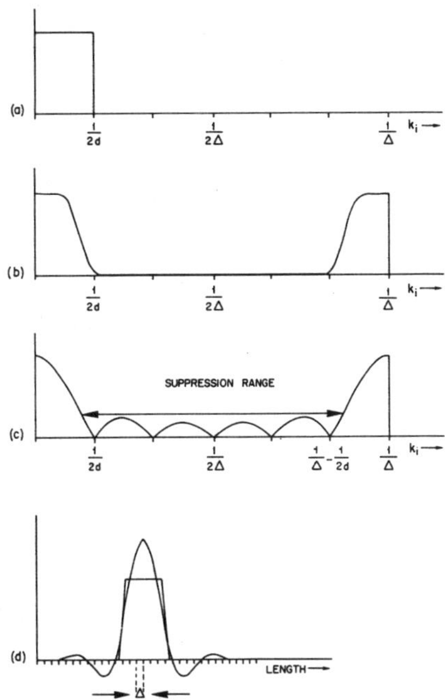

Fig. 4.5. Pattern as digital antialias filter. Δ = distance between pattern elements, d = new sampling interval. In this example $d = 3\Delta$. (a) Desired response, (b) Normal digital antialias filter response, (c) Pattern response with equal filter coefficients, length of pattern is 2d. Suppression range of filter is indicated, (d) Filter coefficients corresponding to (b) and (c) above. Length of digital filter is 31Δ. Length of pattern is 6Δ.

band for $k_{s,r} > 1/2d$, d being the new sampling interval ($=dx_s$ or dx_r, shot or receiver station spacing). Figure 4.5b shows that a digital antialias filter can get very close to this ideal, though it is periodic and has pass areas also for $k_{s,r} = n/\Delta$, Δ being the distance between the filter points. Using a field pattern as a digital antialias filter we can push these pass areas as far out as desired by reducing Δ (see also Section 4.6.5). The patterns used in data acquisition usually have equal weights (= filter coefficients). This leads to the well-known (linear) pattern response as sketched in Figure 4.5c for a pattern with length 2d:

$$p_s(k_s) = \frac{\sin N\pi k_s\Delta}{N \sin \pi k_s\Delta} \quad \text{or} \quad p_r(k_r) = \frac{\sin N\pi k_r\Delta}{N \sin \pi k_r\Delta}, \tag{4.4}$$

for shots and receivers, respectively, N is the number of pattern elements, $N\Delta = 2d$. Patterns with length $N\Delta = 2d$ are called fully overlapping patterns, or patterns with 100 percent overlap, because both ends of the pattern reach to the centers of the neighboring patterns.

Comparison of the response shown in Figure 4.5c with the response in Figure 4.5b shows that the usual field pattern response is very bad from a digital filter point of view: the response (1) is not at all flat in the passband, and (2) has large nonzero values for $k_{s,r} > 1/2d$.

Theoretically, responses like the one in Figure 4.5b can be achieved by giving proper weights to the pattern elements. However, as also illustrated in Figure 4.5d, a reasonable antialias filter has a length at least 9 times that of the new sampling interval, which is rather impractical to implement. Patterns with more conventional weighting functions in which the weights are all positive, e.g., Chebychev weighting (Holzman, 1963) are also quite difficult to lay out correctly in the field.

Another reason why there is not much point to aim for very sophisticated pattern weighting is the limitation caused by sampling noise (see Section 4.3.1) and statics. The problem of sampling noise is extensively discussed in Newman and Mahoney (1973), a classic paper, that shows sampling noise can have drastic effects on the pattern response especially for the more sophisticated weighted patterns. Berni and Roever (1989) provides a detailed discussion on the effect of static variations on the pattern response, illustrated with data from a very densely sampled microspread. The disturbances make the pattern response nonzero phase leading to erratic (spatial) phase shifts of the output traces.

Pattern weights are perhaps easiest to implement in marine streamers, though even there variations in hydrophone sensitivity will impair the pattern response.

The next section discusses that shot patterns should also be applied. However, weighting of shot patterns is, in general, even more difficult to arrange than weighting of receiver patterns.

The accuracy of weighting is irrelevant once the seismic line is shot and recorded at the basic sampling interval with single or clustered geophones. Then any variation in geophone-to-ground coupling and in shot strength can be tackled in the computer.

4.6.2 Two-dimensional pattern response

Receiver *and* shot patterns are required to reduce aliasing whenever larger than basic sampling intervals are used. Figure 4.6 describes the combined use of shot and receiver patterns both in (x_s, x_r) and in (x_m, x_o). The convolution of the two patterns leads to a two-dimensional box-car filter in (x_s, x_r). The corresponding wavenumber response should be described in terms of k_s and k_r and not just k.

Figures 4.6a and c clearly show that patterns mix structural (x_m-dependent) as well as moveout (x_o-dependent) information.

For linear patterns of N elements spaced at Δ_s and Δ_r, respectively, i.e., length of pattern is $N\Delta_{s,r}$, the combined response is

$$p_s(k_s) \times p_r(k_r) = \frac{\sin N\pi k_s\Delta_s}{N \sin \pi k_s\Delta_s} \times \frac{\sin N\pi k_r\Delta_r}{N \sin \pi k_r\Delta_r}. \tag{4.5a}$$

In the (k_s, k_r) plane the zero-crossings of the patterns are represented by lines $k_{s,r} = n/N\Delta_{s,r}$ $(n \neq 0)$. The zero-crossings are illustrated in Figure 4.7a for $\Delta_s = \Delta_r = \Delta$. Only the cross-hatched square in the center represents the desired pass area of the two patterns.

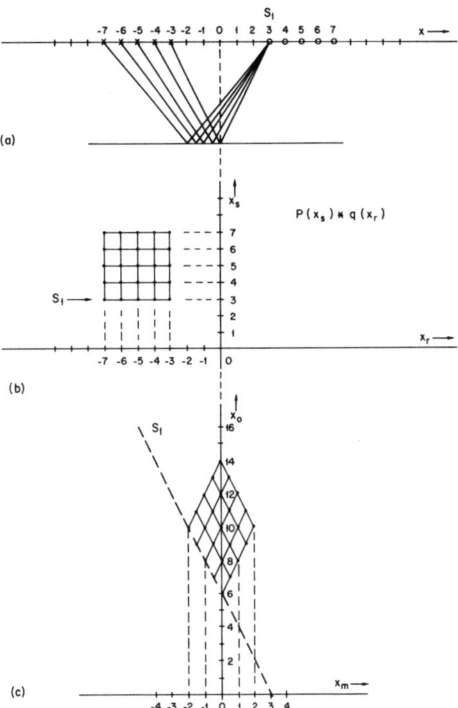

Fig. 4.6. Shot and receiver pattern in (x_s,x_r) and in (x_m,x_o) coordinate systems. (a) 5 shots into 5 receivers, S_1 is at $x = 3$, (b) All elements of this two-dimensional box-car in (x_s,x_r) are added to form one recorded trace, (c) Diamond-shaped filter in (x_m,x_o) showing that patterns mix structure and moveout.

The combined effect of the two patterns gives reasonable suppression in the larger part of the reject area, except along the $k_{s,r}$ axes, where one of the patterns has its pass area. Figure 4.8a shows an isometric display of the pattern response.

The (k_m,k_o) response of the combination of shot and receiver patterns having an equal number of elements and distance Δ between pattern elements can be found by substitution of equation (3.16) in equation (4.5a):

$$p_s \times p_r = \frac{\sin N\pi(k_m/2 + k_o)\Delta \times \sin N\pi(k_m/2 - k_o)\Delta}{N \sin \pi(k_m/2 + k_o)\Delta \times N \sin \pi(k_m/2 - k_o)\Delta}. \tag{4.5b}$$

The pass area and first zero-crossings of this two-dimensional filter are shown in Figure 4.7b.

Equation (4.5b) can also be written as

$$p_s \times p_r = \frac{\sin^2 N\pi(k_m/2)\Delta - \sin^2 N\pi \, k_o\Delta}{N^2[\sin^2 \pi(k_m/2)\Delta - \sin^2 \pi k_o\Delta]}. \tag{4.5c}$$

The use of shot patterns, receiver patterns, or both affects x_m and x_o.

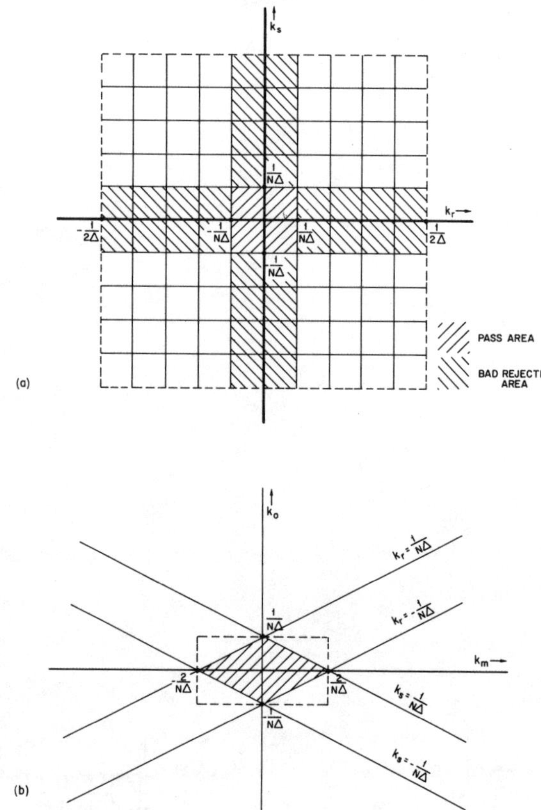

Fig. 4.7. Shot and receiver patterns in double wavenumber coordinate systems. (a) (k_s, k_r) coordinate system, thin lines at $k_{s,r} = n/N\Delta$ ($n \neq 0$) represent zero-crossings. Central square is pass area, (b) (k_m, k_o) coordinate system, first zero-crossings enclose diamond shaped pass area.

Equations (4.5b and c) show that the pattern effect cannot be separated into an action in k_m and one in k_o.

Because it is impossible to describe the action of a shot or receiver pattern separately in x_m or x_o, the discussion is often simplified by projection of the filter points onto the x_o axis. However this projection is valid only for horizontal events (refer to Figure 3.11a). Then the two-dimensional pattern response can be expressed as a function of only one variable. For the real world with varying degrees of x_m dependence of the events (Figure 3.11b) the pattern action in the CMP should not be equated to that in the CRP or CSP.

4.6.3 Overlapping versus adjacent patterns

From Figure 4.5c it follows that overlapping patterns are required to make the first zero-crossing coincide with the Nyquist wavenumber $k_N = 1/2d$. However, the same effect can be reached by using adjacent patterns (having their first zero-crossing at $k_{s,r} = 1/d = 2k_N$) followed

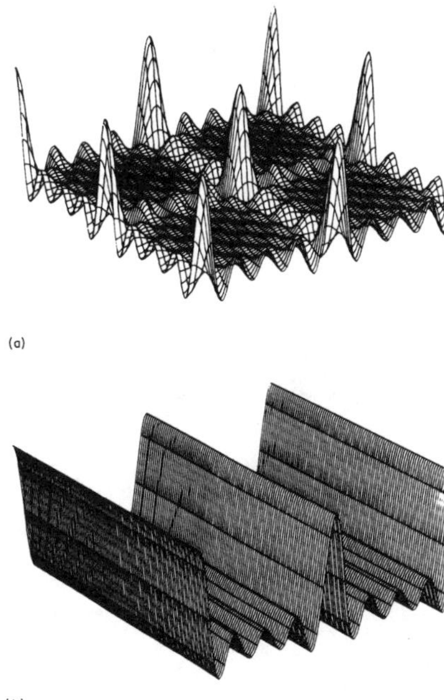

Fig. 4.8. Two-dimensional pattern response. (a) Two patterns with 9 elements each, (b) One pattern with 9 elements. Comparison of (a) and (b) illustrates the need to use both receiver and shot patterns.

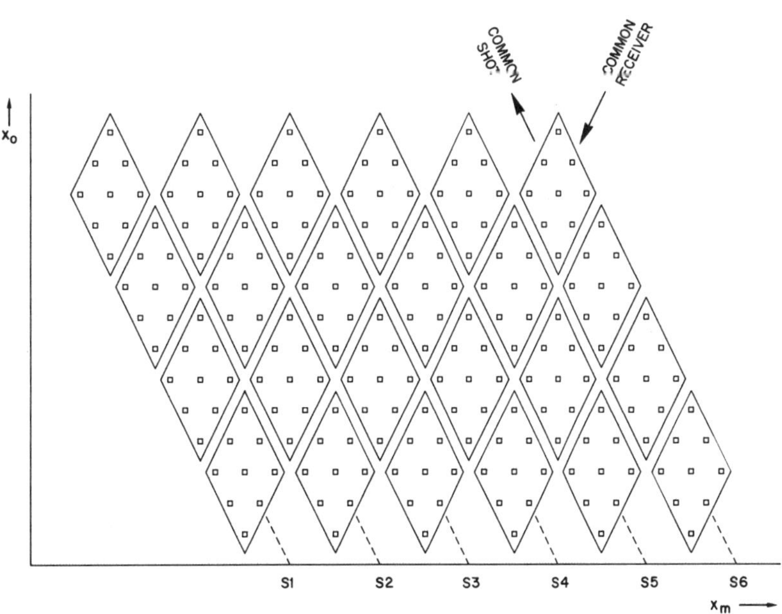

Fig. 4.9. Nonoverlapping shot and receiver patterns in (x_m, x_o). Shot pattern length = receiver pattern length = 3× basic sampling interval. A running 2-dimensional 2 × 2 mix of four diamonds (traces) would create fully overlapping patterns.

by the application of a running two-trace mix of adjacent traces in processing. This process is illustrated in Figure 4.9. Four adjacent traces can be combined into one to form a new trace that would have been recorded with double length source and receiver patterns. Again, not only the longer receiver pattern needs to be simulated; the shot pattern must also be doubled as is discussed further in Section 5.5.

Generally what can be done in the processing center should be done there, provided proper processing principles are pursued. This includes the application of k_s filters in addition to k_r filters.

4.6.4 The effect of not using a shot pattern

If no shot patterns are used, part of the continuous wavefield $W(t,x_s,x_r)$ is not sampled at all. This undersampling is illustrated with the field geometries described in Figure 4.10. The two-dimensional pattern response of a single pattern (9 elements per pattern) is shown in Figure 4.8b. Though all common shots are properly sampled in this case, the total continuous wavefield is still undersampled, so that aliasing exists for the common midpoint, common receiver and common offset panels. Note that the center-spread recording configuration only slightly improves the aliasing problem.

Undersampling means that the original continuous wavefield $W(t,x_s,x_r)$ cannot be reconstructed fully from the sampled values. Wave-equation trace interpolation as proposed in Ronen (1987) aims at reconstructing the zero-offset section $Z(t,x_m)$, not $W(t,x_s,x_r)$, from an undersampled data set. It exploits the redundancy that exists in the data if a model is assumed that consists of primary reflections only. Though this technique may be very useful to create a better stack from undersampled data it cannot fully reconstruct $W(t,x_s,x_r)$ as it does not take surface waves, multiples, and amplitude-versus-offset variations into account.

4.6.5 Distance between pattern elements

The absolute value of the response of a linear pattern with distance Δ between the pattern elements has a periodicity of $k_{s,r} = 1/\Delta$ [cf. equation (4.5a) and Figure 4.5c]. The aim of the pattern is to suppress energy in the range

$$\frac{1}{2d} < |k_{s,r}| \leq |k_{s,r}|_{\max} . \tag{4.6}$$

As can be seen from Figure 4.5c, the suppression range of the pattern is somewhat larger than the range given by

$$\frac{1}{2d} < |k_{s,r}| < \frac{1}{\Delta} - \frac{1}{2d} . \tag{4.7}$$

In general, an antialias filter [$A(t)$ in equation (4.1)] will take care of suppression of the wavefield around and above $k_{s,r} = |k_{s,r}|_{\max}$ (Section 4.2). For long patterns inequality (4.7) may then be replaced by

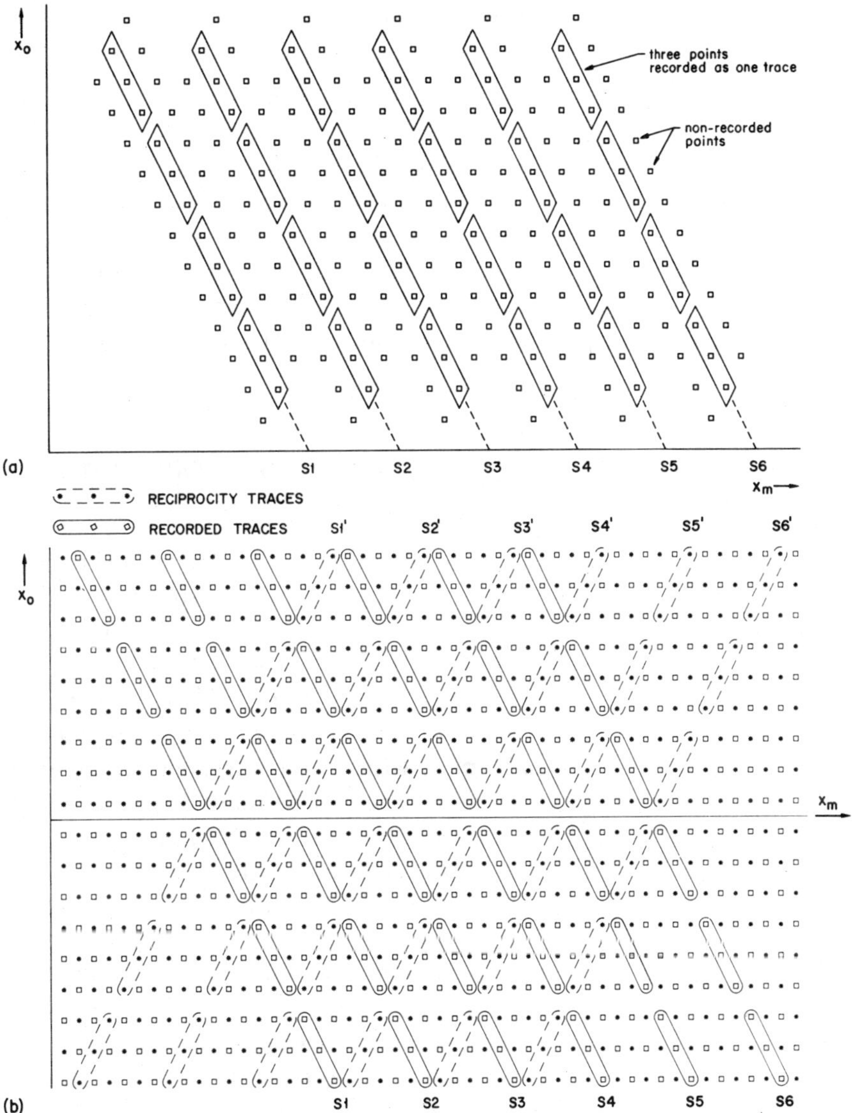

Fig. 4.10. Nonoverlapping receiver patterns and no shot pattern. Receiver pattern length = 3× basic sampling interval. In this case a large part of the three-dimensional wavefield is not sampled. (a) Off-end shooting geometry, (b) Center-spread recording with shot stations halfway between receiver stations. In both cases aliasing occurs in CMP, COP, and CRP.

$$\frac{1}{2d} < |k_{s,r}| < \frac{1}{\Delta}. \tag{4.8}$$

Comparison of inequality (4.8) with inequality (4.6) shows that the distance Δ between pattern elements should satisfy

$$\Delta < \frac{1}{|k_{s,r}|_{\max}},$$

or, with equation (4.3):

$$\Delta < 2\Delta x_{s,r}. \tag{4.9}$$

For small sampling intervals (i.e., shot and receiver spacings) that are still larger than $\Delta x_{s,r}$ it is better to use $\Delta = \Delta x_{s,r}$ in order to improve the suppression between $k = k_N$ and $k = k_{\max}$.

The use of more pattern elements than strictly needed to satisfy inequality (4.9) may be advisable to reduce the effect of variations in geophone-to-ground coupling (see discussion on sampling irregularities in Section 4.3.1) and ambient noise effects. Except for the ambient noise there is not much point, however, in reducing the spacing between pattern elements, we may just as well have a group of phones at each nominal sampling point.

4.6.6 Detrimental effect of patterns on signal

If the signal is aliased in the CMP, correction for moveout will dealias the signal (Section 5.10). Therefore, patterns are not really required to prevent the signal from aliasing in the CMP. In fact, patterns have a detrimental effect on the signal in this domain, in particular reducing the higher frequencies at larger offsets. If $f_{\max} = 75$ Hz and $V_{\min} = 1500$ m/s, then $k_{\max} = 1/20$ m^{-1} and the basic sampling interval is 10 m (Figure 4.11). For a shot and receiver sampling interval of 25 m $k_N = 1/50$ m^{-1}, which means that all events with apparent velocities smaller than 3750 m/s will be more or less affected by the low k_N. Figure 4.12 shows the effect on a wavelet corresponding to a 35 Hz Ricker wavelet using 50 m shot and receiver patterns. [As a rule, Ricker wavelets are not recommended for modeling studies (Hoskin, 1988). However, the Ricker wavelet is very convenient for computing composite responses, as needed here.] The effect is a reduction in amplitude for larger offsets and a broadening of the main loop. Owing to initial blanking, the range of offsets would normally be restricted for $T_0 \leq 1200$ ms. A field experiment demonstrating the effect of shot patterns on reflections with low apparent velocities is described in Brummit (1989).

With large multiplicities and a long spread (i.e., large window in x_o), the stack response will have a very strong noise suppression effect. Thus, in some situations the patterns are required only to prevent aliasing in the x_m coordinate.

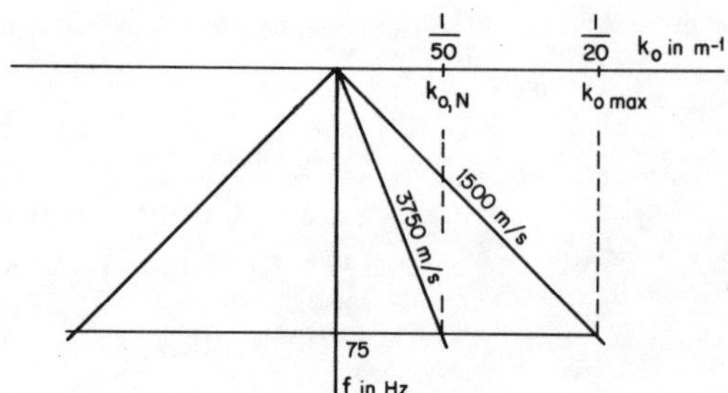

Fig. 4.11. Effect of pattern in (f,k_o). Events with apparent velocities smaller than 3750 m/s will be affected below 75 Hz, if sampling interval in x_o is 25 m.

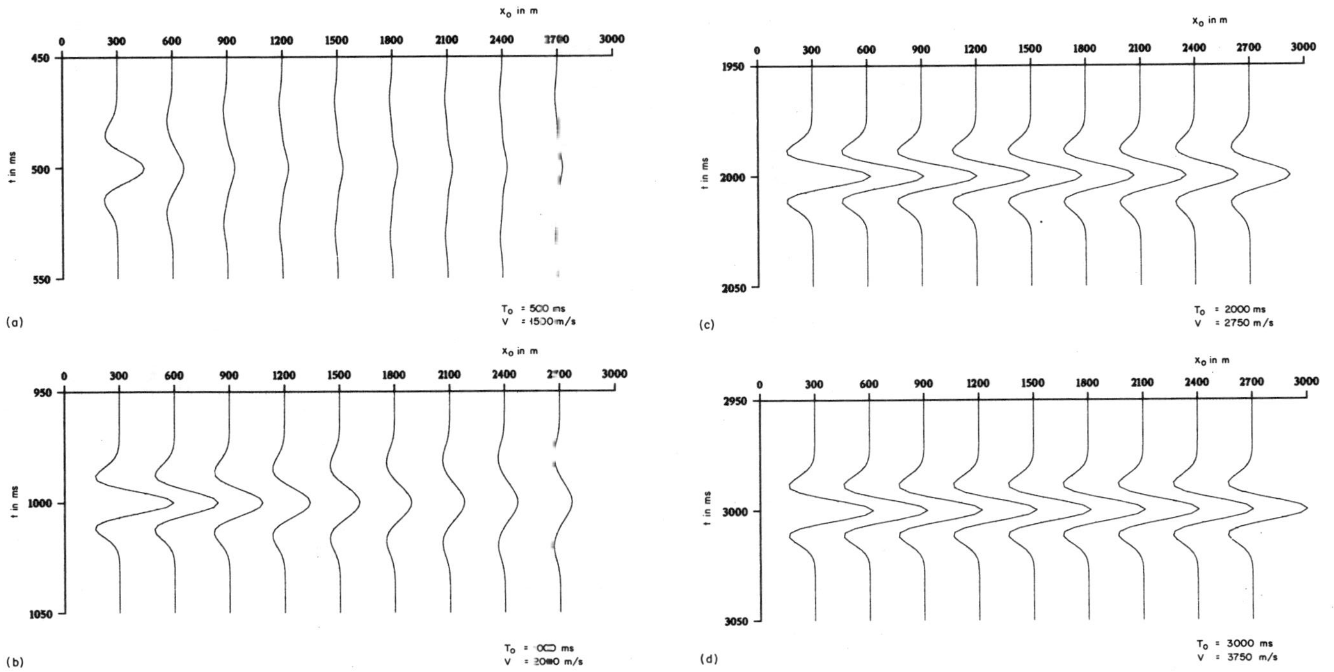

Fig. 4.12. Effect of 50 m shot and receiver patterns on amplitude as a function of offset for a 35 Hz Ricker wavelet for different rms velocities and normal incidence times. (All events have been shifted to a constant time level for easier comparison.) (a) $T_0 = 500$ ms, $V = 1500$ m/s, (b) $T_0 = 1000$ ms, $V = 2000$ m/s, (c) $T_0 = 2000$ ms, $V = 2750$ m/s, (d) $T_0 = 3000$ ms, $V = 3750$ m/s.

4.6.7 Need for patterns to suppress signal

It is customary and often adequate to consider the largest dip of interest when deciding on the required sampling interval. Figure 3.18 illustrates that, for a large part, a diffraction consists of steep flanks in (t, x_m). So, when there are many diffractions, the wavefield in (t, x_m) will show many steep events with velocities close to the minimum apparent (signal) velocity. Close to large faults, salt dome boundaries, and thrust zones the diffraction energy may entirely obscure the events of interest. If the diffractions are properly sampled, much of their energy can be removed by (f, k_m) filtering (next to suppression by stacking), or can be focused by migration. If the sampling interval is larger than the basic signal sampling interval, patterns may also be required to suppress diffraction energy at the expense of the higher frequencies at large offsets.

4.6.8 Symmetric sampling

If shot and station spacing are larger than the basic sampling interval shot and receiver patterns should be used. In order to preserve reciprocity symmetric sampling of the wavefield should be carried out, i.e., not only should shot and receiver interval be equal, shot and receiver patterns should also have equal length and have the same number of elements (see also Sections 3.6.1 and 5.11.6.3). Asymmetric sampling leads to observable artifacts, even if shot and receiver intervals are the same. Artifacts may be visible for off-end shooting and also for center-spread shooting.

Often in off-end shooting, particularly for marine data, parallel lines shot in opposite directions look different, especially for layers with many diffractions. The reason marine lines often show the characteristics of asymmetric sampling is that marine data acquisition using air gun arrays as a source effectively only employs receiver patterns and no shot patterns. The air gun arrays are designed to provide a sharp pulse, but not to provide a lateral antialias filter.

The effect of asymmetric sampling in off-end shooting can be better understood by inspection of Figures 3.18a,b,c, and d. In all four parts of this figure the traveltime curves in the central part are generally closer to horizontal on the right-hand side than on the left-hand side. A receiver pattern does not affect the signal at the right-hand side as much as on the left-hand side. If no shot pattern is used the left-hand parts of the diffraction will be suppressed more than the right-hand parts. Thus on the stack of these data the right-hand side of the apex of a diffraction will be stronger than the left-hand side. On the far right and left the diffraction becomes "two-dimensional"; then there is no longer any difference, as it does not matter for the pattern response whether the signal comes in first at the right-hand or at the left-hand side of the pattern. Therefore, in the stack the diffraction tails will still be symmetric.

In center-spread recording artifacts due to asymmetric sampling may also be observed. As an example consider the CMPs displayed in Figure

4.13. Figure 4.13a shows NMO-corrected surface recordings of Vibroseis data, hence a first requirement for reciprocity has been met: equal depth of shots and receivers. Shot and receiver spacings and receiver pattern lengths were equal to 25 m and a symmetric spread was laid out with maximum offset to both sides of 1537.5 m and nearest offset 62.5 m. The shot pattern length was approximately 30 m, due to the physical size of the three vibrators. Shot stations were half-way between receiver stations in order to benefit from reciprocity as discussed in Section 4.5 (Figure 4.4a). The CMP in Figure 4.13a was sorted according

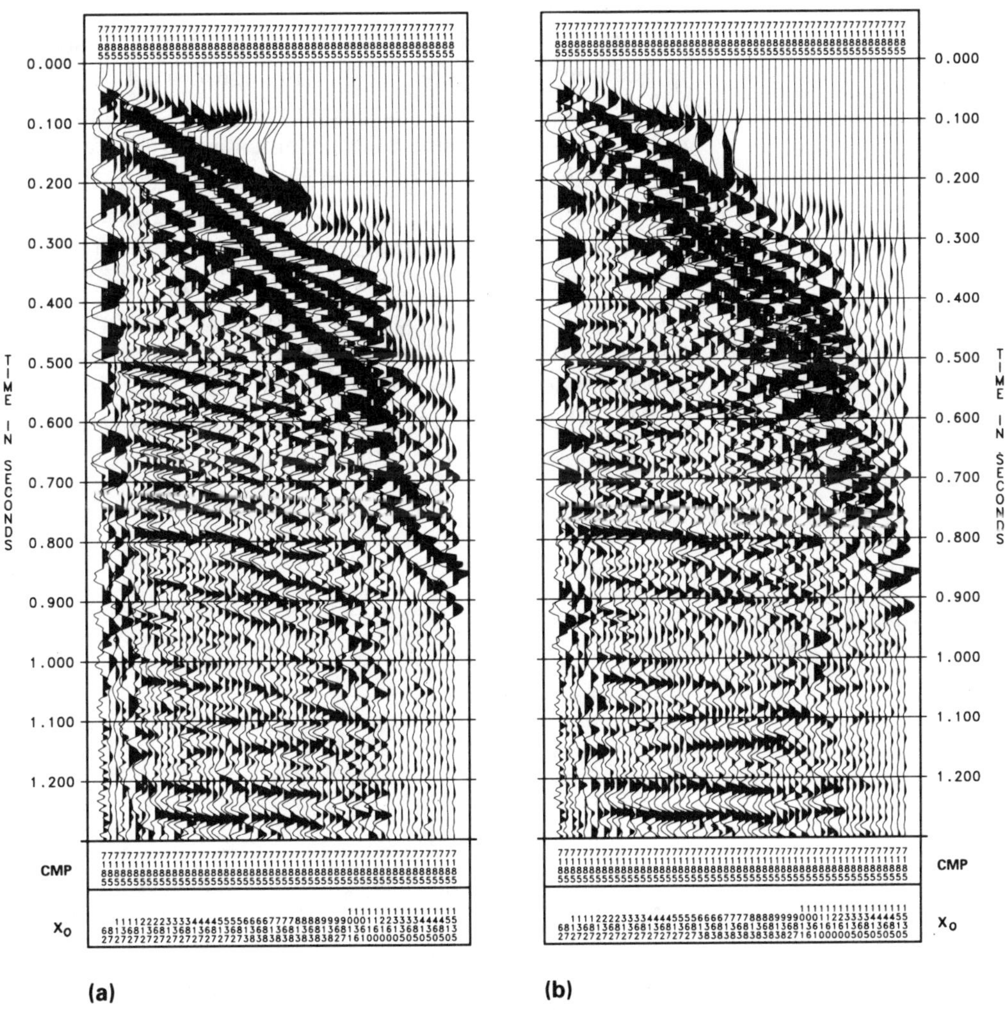

Fig. 4.13. Illustration of odd/even effect in CMP due to asymmetric sampling. (a) CMP recorded center-spread with shot spacing and geophone spacing of 25 m, receiver pattern length of 25 m, and shot pattern length of 30 m. There is hardly any odd/even effect visible in this display. (b) Same data as in (a) after simulation of a 75 m geophone pattern length. A strong odd/even effect is visible at nearly all levels. The event at 1210 ms is a reflection from a horizontal layer and does not show a clear odd/even effect. For explanation see text and Figure 4.14.

to increasing absolute offset, hence traces with positive and negative x_o alternate (not counting some missing traces).

Though Figure 4.13a data are rather noisy the overall impression is one of continuity, where there is no obvious difference between traces from the right- or the left-hand side of the spread. Even though shot and receiver patterns were not exactly equal, reciprocity still applies approximately.

The same data were used to simulate the effect of using very different pattern lengths. A running three-trace mix was applied to the shot records to simulate 75 m geophone patterns while keeping the geophone station spacing at 25 m. After sorting and NMO the CMP corresponding to Figure 4.13a is displayed in Figure 4.13b. A clear odd/even effect is now visible.

Figure 4.14a illustrates that reciprocity is not preserved for the shooting geometry of Figure 4.13b. (In Figure 4.14 the shot pattern length is chosen as 25 m and the geophone pattern length as 75 m.) Trace P (x_{s_P}, x_{r_P}) is a summation of all elementary traces covered by rectangle ABCD with short side of 25 m parallel to the x_s axis and long side of 75 m parallel to the x_r axis. The "reciprocity trace" $Q(x_{s_Q}, x_{r_Q})$ with $x_{s_Q} = x_{r_P}$ and $x_{r_Q} = x_{s_P}$ is the summation of all elementary traces covered by rectangle A$'$B$'$C$'$D$'$. This rectangle is the mirror image with respect to $x_o = 0$ of rectangle ABCD. If this trace were recorded, it would be reciprocal to the trace at P. However, recorded traces have geophone pattern length of 75 m, regardless of their position in the spread (e.g., EFGH), whereas the reciprocity trace has shot pattern length of 75 m and geophone pattern length of 25 m.

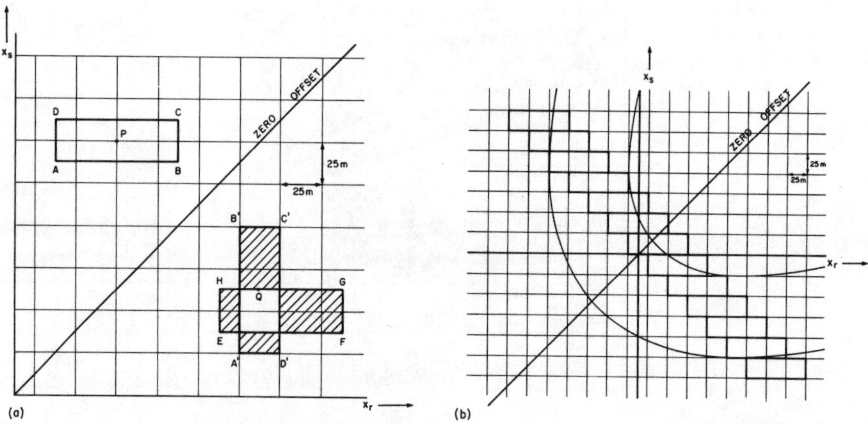

Fig. 4.14. Asymmetric sampling in center-spread recording. (a) Rectangle ABCD represents the convolution of a shot pattern of 25 m with a receiver pattern of 75 m. ABCD is reciprocal with A$'$B$'$C$'$D$'$, but not with EFGH. (b) Time contours of a dipping event illustrate that asymmetric two-dimensional patterns have a different effect on the signal on opposite sides of $x_o = 0$, whereas the effect is similar for patterns on the same side of $x_o = 0$.

For a further explanation of the odd/even effect in the CMP a few common time contours of a dipping event are shown in Figure 4.14b (cf. Figure 3.20). The two-dimensional patterns of neighboring traces in one CMP are indicated schematically. All patterns in the $x_s > x_r$ region cross the time contours at a similar angle, whereas the patterns in the $x_s < x_r$ region cross the time contours at a different angle. Hence the averaging effect of the patterns is different for positive and negative offsets, leading to a difference in character between $x_s > x_r$ traces and $x_s < x_r$ traces. There is no odd/even effect for events that are horizontal in the COPs. Such an event is present at 1210 ms in Figure 4.13.

The observations in this section lead to some important conclusions. Asymmetric sampling in off-end shooting leads to asymmetric diffractions in the stack. Asymmetric sampling in center-spread shooting does not lead to asymmetry in the stack, but to misalignment in the CMP of events with dip in the COP. The misalignment leads to destruction of higher frequencies in the stack.

A frequently used asymmetric recording technique is the shooting geometry with equal shot and receiver station spacings (larger than the basic sampling interval), with adjacent or overlapping geophone patterns, but without a shot pattern, e.g., single hole dynamite. In structurally complex areas the odd/even effect in the CMP may significantly degrade the stack.

Note that a systematic positioning error of the shot in center-spread recording also leads to an odd/even effect in the CMP. This effect increases with increasing moveout, whereas the odd/even effect due to asymmetric sampling increases with increasing structural dip.

4.6.9 Up-dip shooting versus down-dip shooting

Consider the suggestion to shoot marine data up-dip (sail down-dip) in order to reduce the effect of the receiver pattern on the reflections of dipping interfaces. In Figure 3.20c the CSP and CRP trajectories are drawn for $x = 2000$ m and for the case that the boat sails down-dip. This situation also corresponds to Figure 3.19 with $x_s > x_r$. The trajectories for CSP and CRP would be interchanged for down-dip shooting with $x_s < x_r$.

Figure 3.20c shows that in the case of up-dip shooting the CSP sees events that are not as steep as in the CRP. Hence, for up-dip shooting, the receiver patterns do not suppress the signal as much as for down-dip shooting. Therefore the effectiveness of the patterns as antialias filters is less positive when shooting up-dip than when shooting down-dip.

If only gentle dips exist, the interval required for proper sampling of the reflection events in the COP can be quite large (Figure 3.19f), perhaps even larger than the actual interval resulting from the chosen shot and receiver spacings dx_s and dx_r. The NMO-correction would remove aliasing that might exist in (t,x_o) for large dx_s and dx_r. Hence patterns are not required as lateral antialias filters in the case of gentle dip (assuming there is no noise and no diffractions), and patterns do least harm if shooting is up-dip.

However, for steep dips lateral antialias filtering may be needed to prevent aliasing in the (t,x_m) subspaces. Then shooting down-dip ensures the most effective antialias action of the receiver pattern.

The objection, that it is better to record aliased signal than no signal at all, may be true of single, strong events that can be followed easily. However, it is not true of whole packages of steeply dipping strata, of weak strongly dipping sedimentary features, and of fault planes. In these situations there may be so much interference of the various events with each other that aliasing makes them indiscernible. If the data are to be migrated, artifacts result when the reflection events are aliased.

Of course, the dilemma discussed in this section is one of choosing the lesser of two evils, the dilemma dissipates entirely if symmetric sampling is carried out.

4.7 Spatial Windows

The spread length and the position of the shot with respect to the spread determine the window $B[x_o]$ [equation (4.1)] of offsets covered. As stated in Section 3.2, the zero-offset panel containing the normal incidence response of one wave type is often a desirable intermediate product of seismic processing. The reconstruction of this zero-offset section is facilitated if there is no gap in the offset range around $x_o = 0$.

The largest offset to be recorded depends on many different requirements. Optimum velocity determination usually requires large offsets (Schultz et al., 1983). In complicated geologies with nonhyperbolic moveout effects large offsets can be very diagnostic but require much extra effort in processing. For a given budget and in such areas use of the basic signal sampling interval with a short offset range rather than recording with larger (shot and receiver) intervals and a longer offset range may be better.

The other spatial window $B[x_m]$, line length, depends mostly on the exploration target. However, considerations of full multiplicity and requirements for migration operator length also play a role.

4.8 Ghosts

The ability to reconstruct acoustic impedance also for very low frequencies depends largely on the ghost effect. Therefore, some extra attention is paid to this part of the earth response.

In marine data acquisition with $z_s = d_s$ and $z_r = d_r$ each reflection is followed by three ghosts caused by reflection at the surface of part of the energy (Figures 4.15a and b). Disregarding the direct arrival, the total wavefield can then be approximated by

$$W(t,x_s,x_r) = G_s(t,x_s,x_r) * G_r(t,x_s,x_r) * W_d(t,x_s,x_r), \qquad (4.10a)$$

with

$$G_{s,r}(t,x_s,x_r) = 1 + r \exp(-j\omega\tau_{s,r}). \tag{4.10b}$$

Here G_s and G_r are the source and receiver ghost filter respectively, τ_s and τ_r are the corresponding time delays, r is the reflection coefficient at the surface, and W_d is the (desired) wavefield that is not affected by the ghosts.

The time delays τ_s and τ_r are functions of x_o and t and not of x_m as long as d_s and d_r remain constant.

As can be seen from equation (4.10b) the ghost filter changes phase and amplitude of the wavefield. For small offsets the extra travel path is near vertical. Then the time delays may be approximated by

$$\tau_{s,r} = \frac{2d_{s,r}}{V_w}. \tag{4.11}$$

For $r = -1$, equation (4.10b) reduces to a filter with zeroes at $f = nV_w/2d_{s,r}$, $n = 0, \pm1, \pm2, \ldots$ (Figure 4.15c). Hence the depths of source and receiver have a critical effect on the bandwidth of the recorded wavefield. The shallower the source and receivers the more the ghost effect will resemble that of a differentiator, thus reducing the ability to record low frequencies. The deeper the source and receivers the lower will be the frequency f_g of the first notch of the ghost filter. Therefore, the best compromise may be to select d_s and d_r such that f_g coincides with the maximum usable frequency $f_{\max}$ (for $x_o = 0$), i.e.,

$$d_s = d_r = V_w/2f_{\max}.$$

For land data acquisition using a buried source, the ghost effect can be much more complicated, because the energy may bounce back from several levels above the source. On the other hand, a surface source such as Vibroseis or weight dropper has the advantage of not creating a ghost signal.

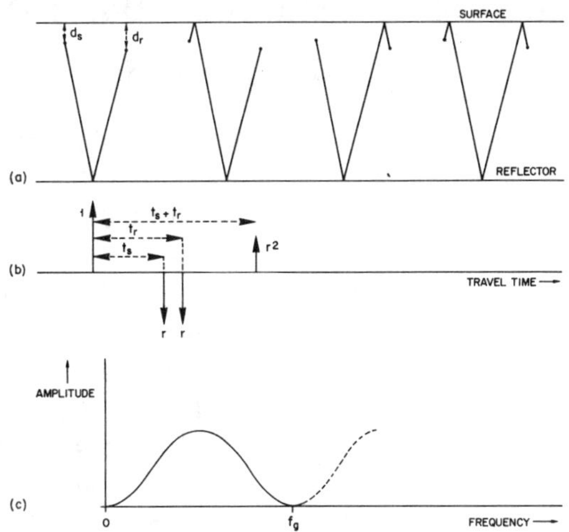

Fig. 4.15. The ghosting effect. (a) Multiple reflection paths, (b) Primary reflection and ghosts, (c) Effect on amplitude spectrum of one ghost ($r = -1$).

4.9 Resolution

This text does not repeat considerations on what resolution is achievable for a given sampling of the continuous wavefield. An extensive treatment of seismic resolution is given in Berkhout (1984). We concentrate on the data acquisition parameters required to achieve maximum resolution.

Figure 3.8b clearly shows that for any given maximum frequency the full wavefield can be reconstructed for $f < f_{max}$ if the basic sampling interval is used, i.e., $k_N = k_{max}$, where k_{max} is given by equation (3.20). Finer spatial sampling means oversampling; coarser spatial sampling leads to a reduced horizontal resolution. Usually we are interested only in a particular wave type (the signal) for which the minimum V_{app} is larger than V_{min}. The above reasoning is also valid for this wave type, using the basic *signal* sampling interval. However, finer sampling may lead to better potential to remove unwanted wave types with a lower minimum V_{app}.

I propose to call the recording technique in which the basic signal sampling interval is used *full-resolution recording*.

As observed earlier, in conventional data acquisition, there is usually a large difference between the wavenumber spectrum and the frequency spectrum at their low ends because very low frequencies, below 8 Hz, generally are not recorded. The absence of low frequencies may be due to inadequate impulse response of geophones or hydrophones, absence of low frequencies in the generated source signature, and/or the ghosting effect. Hence, even with full-resolution recording the ideal of equal spatial and temporal resolution is still not reached. This lower limit to the bandwidth of the seismic data forms a serious obstacle in the reconstruction of the acoustic impedance.

4.10 Illustrating Some Basic Principles

4.10.1 Microspreads

Recording the continuous wavefield is normally done with sampling intervals that are too large to allow an analysis of the wavenumber spectra of the continuous wavefield. Patterns remove most of the short wavelength components of the wavefield. For a better understanding of the geophysical problem in an area, it can be very helpful to shoot and analyze *microspreads* (also called "noise tests" or "noisespreads"). A microspread is "A spread with very short geophone group intervals (1 to 15 ft), used in *noise analysis*" (Sheriff, 1984).

Figure 4.16 shows two microspreads shot in the same area in The Netherlands. The source was Vibroseis, the receivers were spaced 2.5 m apart. For one microspread the source remained at the same location with 40 channels per shot (Figure 4.16a), for the other one the spread (30 receivers) remained at the same location and the source was moved between shots (Figure 4.16b). The inserts show the shooting configurations in (x_m, x_o).

In both figures Vibroseis correlation noise can be observed at shallow levels (running parallel to the dominant groundroll event).

Figure 4.16a illustrates that the groundroll generated by Vibroseis is very continuous. Even though separate sweeps were used for every 40 receivers, discontinuities at the changeover points from one shot to the next are hardly visible. This continuity illustrates the stability of Vibroseis with its low energy density, not causing any deformation at the shotpoint. A similar microspread shot with dynamite would show discontinuities at the changeover points caused by nonelastic deformation of the shothole.

Figure 4.16b shows a number of discontinuities in the recorded wavefield corresponding to the discontinuities in the shooting configuration as shown in the insert. These discontinuities in the signal illustrate that the groundroll varies with the shot position, i.e., it is x_o and x_m dependent.

At traveltimes close to the first arrival of compressional waves characteristic patterns in the recorded data come back for consecutive shots in Figure 4.16b. This phenomenon is indicative of surface-consistent effects at the spread location.

Section 3.8.2 describes that back scatter has the same slope as the direct arrival with opposite sign. Back scatter of the groundroll can be observed in Figure 4.16a as well as in Figure 4.16b.

The point to point variation of the statics can be analyzed using microspreads (Berni and Roever, 1989). If statics vary rapidly over the length of the source or receiver pattern, then the pattern could cause severe signal distortion. Figure 4.16 shows that statics are not a great cause for concern at that particular location.

Another application of microspreads is the analysis of geophone sensitivity (Section 4.3.1). Variations in geophone sensitivity and coupling may lead to an inadequate performance of source and receiver patterns as antialias filters (Ongkiehong and Huizer, 1987, Berni and Roever, 1989).

An (f,k) analysis of a microspread as in Figure 4.16b would show much spreading of energy over the whole wavenumber range, due to the discontinuities in the recording. However, an (f,k) analysis of a microspread recorded with the shot at a fixed location as in Figure 4.16a shows the relative importance of various wave types and the energy distribution in (f,k_r) space. If the sampling interval of the microspread is smaller than the basic sampling interval, the (f,k_r) analysis can be used to determine the basic sampling interval. As mentioned in Section 4.2, it is possible that waves with the smallest apparent velocity $V_{\min}$ carry little energy for frequencies above f_1, with $f_1 < f_{\max}$, due to strong attenuation for that wave type. In such cases [cf. equation (4.3)]

$$|k_r|_{\max} = f_1/V_{\min} < f_{\max}/V_{\min},$$

so that the basic sampling interval is larger than what follows from $\Delta x_{s,r} = V_{\min}/f_{\max}$.

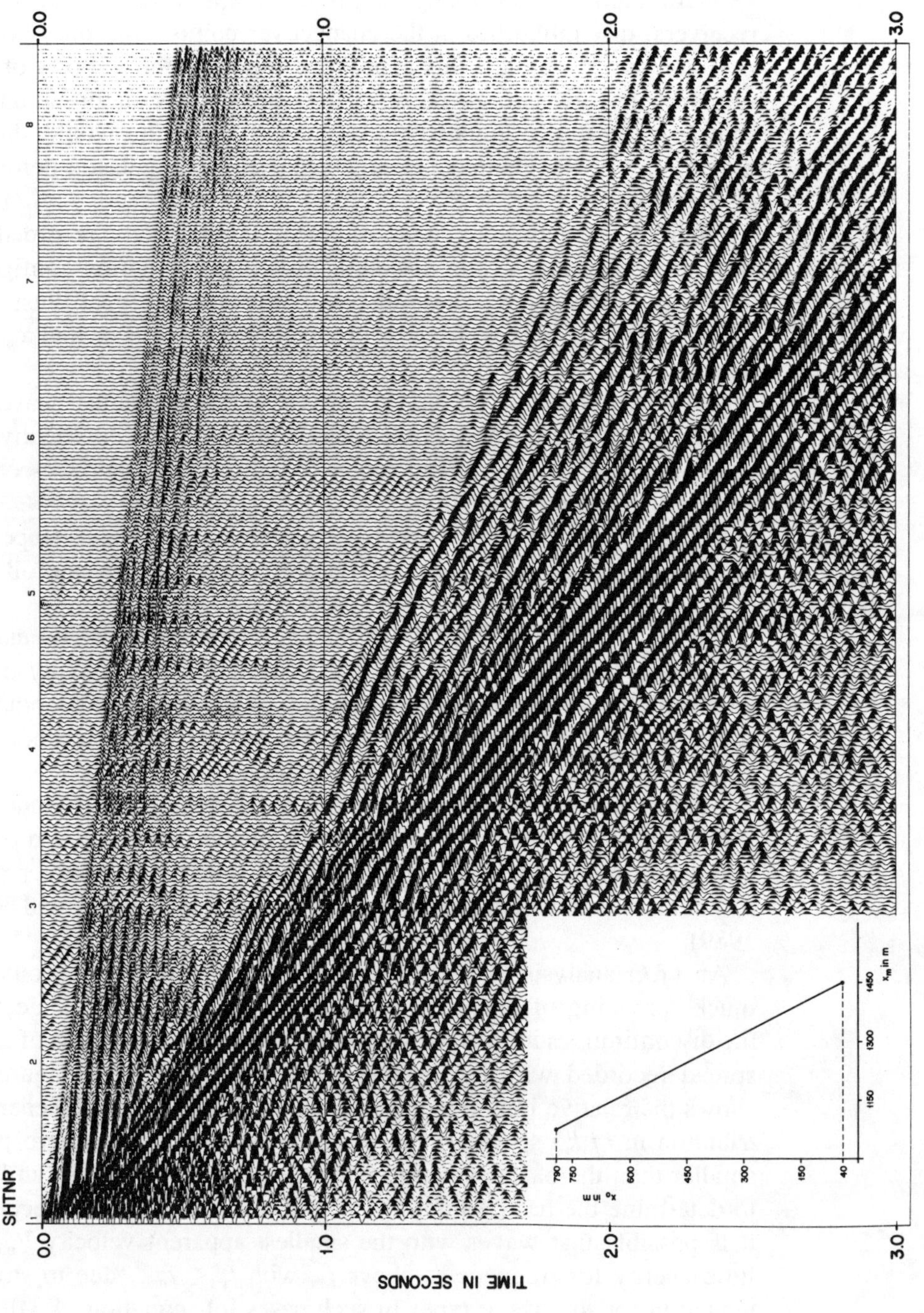

Fig. 4.16. Two Vibroseis microspreads. Distance between receiver positions: 2.5 m. Inserts show shooting configuration in (x_m, x_o). (a) Moving spread with 40 receivers per shot. Sweep 5–125 Hz in 10 s,

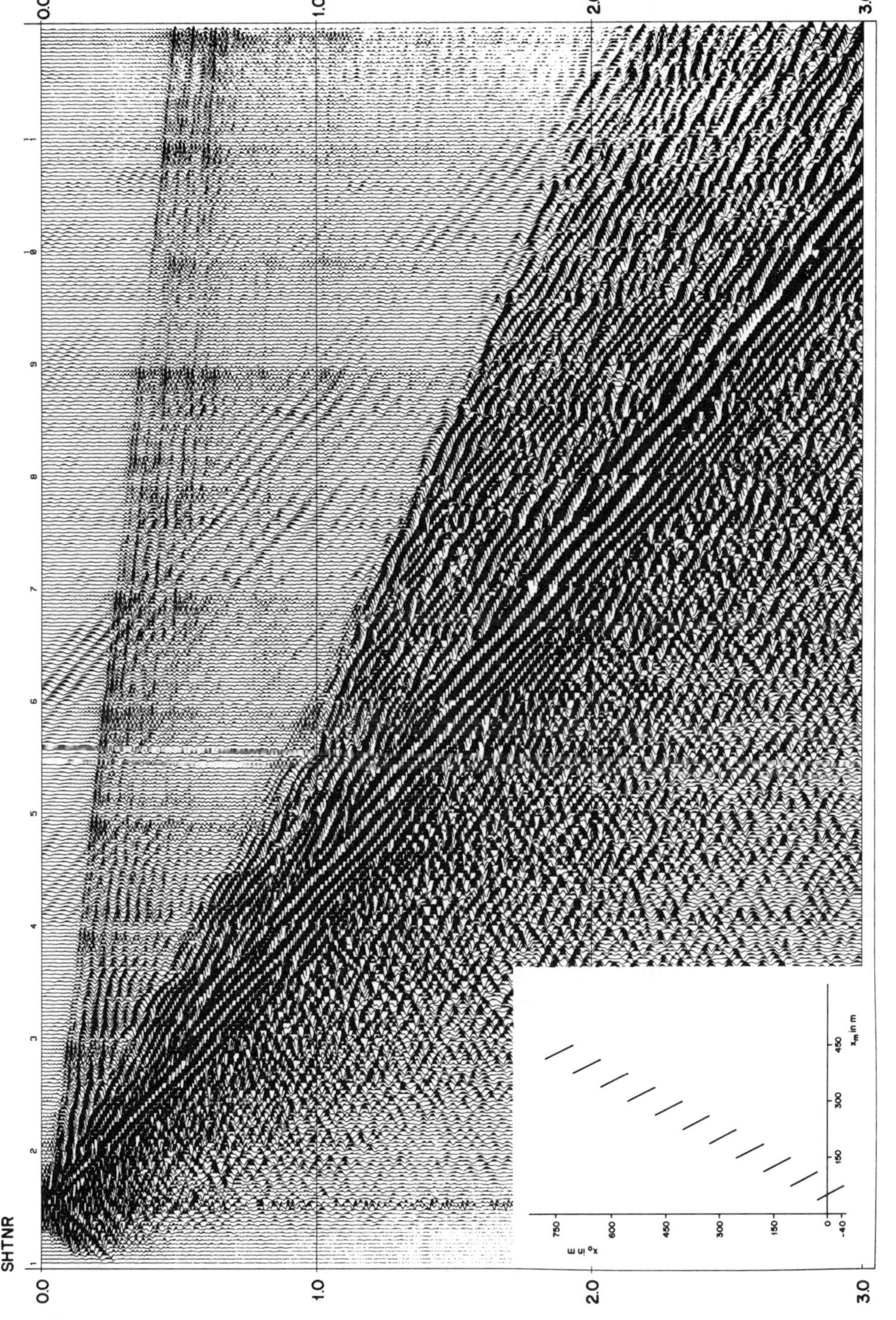

Fig. 4.16. cont. (b) Moving shot with 30 receivers per shot. Sweep 5–77 Hz in 12 s. Discontinuities in data correspond to discontinuities in configuration as shown in insert.

In conventional common shot panels recorded with geophone patterns the groundroll shows up as aliased energy (Figure 1.1). Yet, it may be possible to get a reasonable idea of the minimum apparent velocity present in the wavefield by analyzing the (f,k) spectrum for low frequencies. It is not possible, however, to make an accurate measurement of f_1, the maximum frequency with groundroll energy, because of aliasing. As shown, the microspread is a better tool for this analysis.

Because a microspread only measures a local response, selecting a representative location for the microspread is very important. The best location for shooting a microspread may be found by inspecting shot records of an existing line in the area.

Microspreads (with fixed shot location) only provide the (f,k) spectrum in one cross-section of the three-dimensional wavefield. By reciprocity, they also show what happens in the common receiver panel. However, because microspreads do not show the spread of energy over the (k_m,k_o) spectrum, they do not allow measurement of the midpoint dependence of signal and noise. Yet, some idea about the total amount of energy that is midpoint dependent, may be derived from (f,k_r) analysis of a one-sided $(x_o > 0)$ microspread. Most energy of the microspread will be concentrated in one half of the (f,k_r) plane, say in $k_r > 0$, because traveltime of most events increases with increasing offset. The energy in the other half of the (f,k_r) plane $(k_r < 0)$ is for a large part midpoint-dependent energy. (Another part is caused by statics, sampling noise, and x_o-window truncation noise. Corrections can be made for these effects.) Assuming no preference for right or left dipping events in x_m, the total midpoint dependent energy is twice the energy present in $k_r < 0$.

Knowledge of the full (f,k_m,k_o) spectrum would be required to analyze the effect of a compromise shooting geometry, such as not using a shot pattern.

4.10.2 Independence of spatial coordinates

The three-dimensional continuous wavefield $W(t,x_s,x_r)$ has two independent spatial coordinates x_s and x_r. As a consequence wavenumber filtering in one spatial domain does not produce the same results as applying the same filter in the other spatial domain. This point is discussed in Section 3.6.1 (Figure 3.6), and illustrated in Section 4.6.8 (Figures 4.13 and 4.14).

Another illustration is provided with Figures 1.1–1.4. These figures were briefly mentioned in the Introduction. The data shown in Figure 1.1 were shot according to the *symmetry criterion* (Section 5.11.6.3): geophone patterns with 12 geophones spaced at 2.5 m, and vibrator locations also chosen at every 2.5 m, for a shot and geophone station interval of 30 m. A center-spread configuration was applied. A configuration with equal shot and geophone intervals allows the application of identical k or (f,k) filters in CSPs and in CRPs.

Figure 1.2 shows the same data as Figure 1.1 after application of a (f,k) filter in the CSPs, and Figure 1.3 shows the data of Figure 1.1 after application of the same (f,k) filter in the CRPs. Both filters were applied to the whole data set followed by resorting for display. The panels in which the filters were applied show a clear mixing effect, whereas the other panels do not show any mixing effect. Yet, all traces in the various panels have been modified by the application of the filters. Careful examination of the CMPs shows that the groundroll noise level has been reduced. This noise level reduction in each trace of the CMP has been effected using traces that do not belong to that same CMP.

After application of the (f,k) filter in both CSPs and CRPs, the groundroll is further suppressed in the CMP (Figure 1.4). The CMP should now show some sign of mixing as the neighbors of each trace in the CMP of Figure 1.1 have now contributed to each corresponding output trace in Figure 1.4. For a spatial filter length of 5 traces in CSP and CRP, there would be $5 \times 5 = 25$ traces contributing to each output trace, whereas only 3 of those traces would have come from the same CMP. Hence the mixing in the CMP is very limited and not really visible. The same observations would be possible in a display of any COP. After stack, the stacked section would show similar mixing as the CSPs and CRPs, as follows from the equivalence theorem (Section 5.11.5).

4.11 The Selection of Data Acquisition Parameters

4.11.1 Marine data acquisition

At this point it is appropriate to discuss various data acquisition techniques in relation to an optimal recording of the continuous wavefield $W(t,x_s,x_r)$. However, interest is generally only in a good-quality zero-offset section $Z(t,x_m)$, to be obtained from $W(t,x_s,x_r)$ via a number of processing steps. The redundancy in $W(t,x_s,x_r)$ compared to $Z(t,x_m)$ may relax requirements on the data acquisition parameters. The relation between stacking and the field parameters is discussed in Section 5.11.6.

Based on earlier results, Table 1 lists data acquisition parameters for full-resolution recording of marine data assuming $V_{min} = 1500$ m/s. If the source does not produce frequencies above f_N, and if the ambient noise carries low-frequency energy only, a moderate antialias filter $A(t)$ may be selected. Note that especially for shallow water depth the wavefield may contain events traveling at a much lower velocity than the water layer velocity resulting in V_{min} lower than 1500 m/s.

A major practical problem in conventional marine data acquisition is the ship speed for any f_N larger than about 65 Hz. Table 1 shows that the ship speed should then be less than 4 knots, at which speed single-propeller boats are not properly maneuverable. To keep the vessel maneuverable it should be equipped with a dynamic positioning system. Balancing the streamer at low speeds is also more difficult. This is an area in which the seismic method can be further improved.

Table 1. Parameters for marine full-resolution recording, assuming V_{min} = 1500 m/s. No receiver pattern is to be used and the in-line length of a shot pattern should be 0 m.

Nyquist frequency	125	100	62.5Hz
Time sampling interval	4	5	8 ms
Shotpoint interval	6	7.5	12 m
Receiver interval	6	7.5	12 m
Source-to-first receiver	3	3.75	6 m
Depth of source	6	7.5	12 m
Depth of receivers	6	7.5	12 m
Ship speed (6s between shots)	1	1.25	2 m/s
	=2	2.5	4 knots
Maximum frequency in source	125	100	62.5 Hz

A higher resolution can be achieved by shooting the seismic line twice: once with shallow source and receiver, e.g. at 6 m to recover as high frequencies as possible and once with deep source and receiver, e.g. at 20 m to recover as low frequencies as possible. Even better is to have a deep-towed and a shallow-towed streamer at the same time in combination with a deep and a shallow source pattern. This set-up offers great potential for broadening the bandwidth of the final seismic data. A technique going a long way in this direction was reported in Brink and Svendsen (1987).

Data acquisition parameters are selected to obtain the best-quality seismic for a given budget. Compromises are always necessary. Usually selecting a full-resolution recording technique will not be possible, therefore patterns must be used. Pattern lengths must be equal to shot and receiver interval. Shot patterns are as important as receiver patterns. Using equal receiver and shot patterns at equal depths in 3-D surveys would allow adjacent lines to be shot in opposite directions without the risk of asymmetry effects in the cross-line direction, while leading to a much smaller loss of time during line changes.

There is not much point in using shot patterns larger than the shot interval (unless weighting is applied). Without weighting, patterns have a bad response; it is better to use proper high-cut k-filters or better still (f,k) filters in the processing center. Several authors (Lofthouse and Bennett, 1978, Ursin, 1978, 1983, Roksandić, 1986) report better results when using super-long air gun arrays. However, the comparisons in these articles do not show that the total energy used in the short arrays was the same as in the long arrays. Creating a proper shot pattern over the length of the shot interval using the same total energy (perhaps by spreading air guns sideways) followed by proper processing in the computer should lead to equivalent or better results than using long air gun arrays. However, opinions differ on this matter.

Recording the wavefield with a short receiver interval using a 500- or 1000-channel streamer while not having an equally short shot interval is potentially better than using a receiver interval as large as the shot interval. At least one cross-section in (t,x_s,x_r) is properly sampled in that case (Allen, 1988).

Areal patterns (with a component in the y direction) are not needed in 2-D surveys to prevent aliasing of $W(t,x_s,x_r)$. Areal patterns will serve to suppress some side-swipe energy (energy arriving from outside the vertical plane through the seismic line) or to accommodate the physical size of the source. In 2-D surveys whether areal patterns are to be recommended depends on the relative importance of unwanted side-swipe energy and wanted side-swipe energy. Areal patterns for receivers (in marine data acquisition) are not (yet) possible, so using them only at the source end again leads to asymmetric recording.

For 3-D surveys areal patterns are to be recommended if distance between lines is greater than the basic signal sampling interval. Cross-line processes, notably migration, may suffer from aliasing effects if there are no shot and receiver cross-line patterns.

4.11.2 Land data acquisition

Much of the above discussion is also applicable for land data acquisition. Important differences are

—presence of low-velocity wave types necessitates the use of patterns even if full-resolution parameters are used for the wanted wave type.

—on land it is easier to record individually all shots that are meant to be elements of a pattern, allowing better processing of the data.

—areal shot and receiver patterns are possible.

—center-spread recording with shot move-up equal to receiver station spacing and with shot stations half-way between receiver stations immediately produces properly sampled CMPs and COPs (Figure 4.4a).

If use of only a shot or a receiver pattern is acceptable, under certain circumstances using shot patterns rather than geophone pattterns may be more economical, e.g. for Vibroseis recording in built-up areas with no proper places to plant geophones. Another interesting possibility to consider in this context is to record all shots separately, and apply any required k filtering in the computer. This technique allows better statics computation at the expense of some increase in acquisition and processing costs.

For shear-wave recording the wanted wave type has a minimum apparent velocity that is not very different from that of the surface waves. On the other hand, to achieve resolution comparable to that for compressional waves a smaller f_N suffices. Therefore, if P and S data are recorded along the same line, use of the same source and receiver intervals in both recordings is justified. Because of the lower f_N for shear, the basic sampling interval for shear data is larger, hence the number of pattern elements can be smaller.

Further compromises with respect to the use of patterns can be acceptable in 3-D recording. The migration operator then can be the main attenuator of events not satisfying the velocity model.

Because the choice of a field geometry will always remain something of a compromise, it is good to know what the next best geometry would have produced. Shooting one such line will provide valuable information for any future infill shooting. In many cases performing high-quality preprocessing may be necessary; then proper sampling is a must.

There are situations for which even recording at the basic sampling interval of 2-D data will not produce the desired result. Particularly for very rough near-surface conditions, such as basalt-covered areas, there may be so much scattering of energy that other techniques have to be sought.

Chapter 5
PROCESSING THE RECORDED WAVEFIELD

5.1 Principles of Space-Conscious Processing

Ideally, the seismic processing sequence consists of a series of processes that fully honor the spatial and temporal relationships of the original wavefield and of the recording artifacts. I call this type of processing *space-conscious processing*. It is characterized by three phases in each processing operation:

(1) examination of all data of the line (or even survey) as a function of the spatial coordinates,

(2) analysis of the results of phase (1) and determination of spatially varying parameters for the operation, and

(3) application of the operation.

This approach, which has always been used in the determination of surface-consistent residual statics, is equally important in other processing steps such as editing, equalization, and deconvolution.

In the past, seismic processing methods were strongly influenced by the hardware limitations of the computer. Limited data storage capacities and slow tape speeds led to an emphasis on time-domain processes: one trace in—one trace out. Multichannel operations were accepted only when unavoidable (demultiplexing and sorting). Today operations in various space-time domains are no longer a problem. Yet present-day software and present-day processing techniques are often too much time-domain oriented. The following sections discuss space-conscious processing concepts in more detail for a number of operations. Much more elaborate discussions of seismic processing techniques in general are to be found in Hatton et al., (1986) and in Yilmaz (1987).

The importance of space-conscious processing has grown with (1) the increase in spatial sampling rates, (2) the advent of more sophisticated algorithms, and (3) the need for more sophisticated applications.

Data recorded with a high spatial sampling rate have a relatively low S/N ratio, because the correspondingly short patterns have a low noise attenuation which imposes special requirements on the processing. For instance, deconvolution operators derived from single noisy traces may do more harm than good. The simplest "solution" to beat the low S/N ratio is to simulate patterns in processing, thus defeating the purpose of full-resolution recording. Space-conscious processing is essential for full-resolution data.

To be effective (f,k) filtering, (partial) prestack migration, and (τ,p) processing all require proper sampling. These techniques are sensitive to missing shots and receivers and other irregularities occurring when

77

recording the continuous wavefield. Space-conscious preprocessing is required to make those operations work properly.

If the only objective of the seismic survey is to produce data for structural interpretation, quality requirements are less than when the data are also to be used for detailed seismostratigraphic interpretation or for the prediction of lithology and porefill. Particularly when offset-dependent amplitude effects are to be measured (Backus, 1987) full-resolution recording and space-conscious processing are essential.

Although Chapter 5 emphasizes the need for space-conscious computer processing of the seismic data, the man behind the machine is still essential to judge the results and to decide the best course of action.

5.2 Diagnosis of the Recorded Wavefield

Before any other processing is undertaken it may be useful to check the recorded wavefield for the presence of spatial irregularities. If full-resolution recording has been used, the $(f,k_{s,r})$ diagrams should be considered as a diagnostic tool especially for marine data. We saw in Figure 3.10 that the continuous wavefield does not carry energy for the area defined by $|k_{s,r}/f| > 1/V_{min}$. Spatial irregularities such as variations in shot strength, geophone-to-ground coupling, and noise bursts will smear energy into this region. Proper processing removes this energy. The (f,k_s) diagram is most suited to establish the presence of ambient noise (Figure 4.1b). Aliasing of surface waves, aliasing due to missing shots or receivers, irregular depth level of shots and receivers, and irregularities in the spatial position of shots and receivers will also show up as smearing in the (f,k) diagrams. All these effects can no longer be undone, yet the (f,k) diagram will show the confidence which can be placed in the recorded wavefield.

The $(f,k_{s,r})$ diagram will not be as helpful for data shot with larger spatial sampling intervals, because of the relatively small area where $|k_{s,r}/f| > 1/V_{min}$. More generally applicable is the analysis of spatial irregularities in (x_m,x_o). Energy measured in a window and displayed in an (x_m,x_o) diagram will reveal missing shots and receivers, irregularities in shot and receiver strengths, channel sensitivity variations, and near-surface effects, each of these having its own particular "signature" in (x_m,x_o). An early description of a diagnostic technique, Morgan (1970), uses mapping of wavelets as a function of the spatial coordinates to diagnose static time shifts and amplitude variations.

5.3 Editing and Equalization

Editing is an early step in the processing sequence concerned with spotting and correcting noisy traces or trace parts. Editing of land data, which used to be carried out by visual inspection of the recorded data, is now becoming more and more problematic because of increasing volumes of data and decreasing S/N ratios in individual traces. Marine data are often edited only after inspection of the brute stack. Hence, an automated editing procedure that exploits the spatial relationships of the data is very desirable.

To determine whether a trace has exceptionally low or high sample values, statistics must be gathered for the whole recorded wavefield. Thereafter the measurements on each trace must be compared with the statistics valid in its environment. A trace identified as faulty must be flagged in order to prevent its use in the subsequent equalization process. After equalization the trace must be replaced by interpolation in the CSP or CRP.

Equalization is required to correct for source strength variations and geophone to ground coupling variations (for a discussion of geophone-to-ground coupling see Krohn, 1984). On marine data hydrophone sensitivity may vary. Thus, for land data surface-consistent techniques are required to correct for variations (Taner and Koehler, 1981), whereas for marine data a separation must be made between common shot and common hydrophone (i.e., channel-consistent) effects.

Often "equalization" is one of the processing steps appearing in the label of a stacked section. However, equalization usually means normalization to an equal output energy. Normalization of single traces may do more harm than good, particularly for low S/N ratios where the signal may not regain its spatial continuity that existed prior to recording.

Equalization requires similar statistics to editing, and therefore combining the two operations into one process is efficient. Thus, space-conscious editing and equalization consist of the following phases:

(1) Go through all data and gather statistics (energy measurements).
(2) Analyze the statistics, flag faulty traces, then determine shot and receiver strength correction factors.
(3) Go through all data again, apply equalization, and interpolate flagged traces.

5.4 (f,k) filtering

For the basic principles of two-dimensional or (f,k) filtering the reader is referred to Clement (1973). Apparent velocity filtering (also called dip filtering or fan filtering) is the most commonly encountered 2-D filtering in seismic processing, and was introduced in Fail and Grau (1963) and Embree et al. (1963). Cassano and Rocca (1974) discuss the design of (f,k) filters with a minimal mixing effect, based on a data-adaptive approach. Marschall (1980) expands one-dimensional recursive filtering techniques to two dimensions.

In this section an extensive discussion of (f,k) filtering in various domains helps to illustrate once again some aspects of the three-dimensional nature of the wavefield of the 2-D seismic line. As an introduction let us first consider a few records from the Dutch North Sea (Figure 5.1). The first arrival consists mainly of refractions from near-water-bottom interfaces and is followed by a dispersive dangler that broadens with increasing offset.

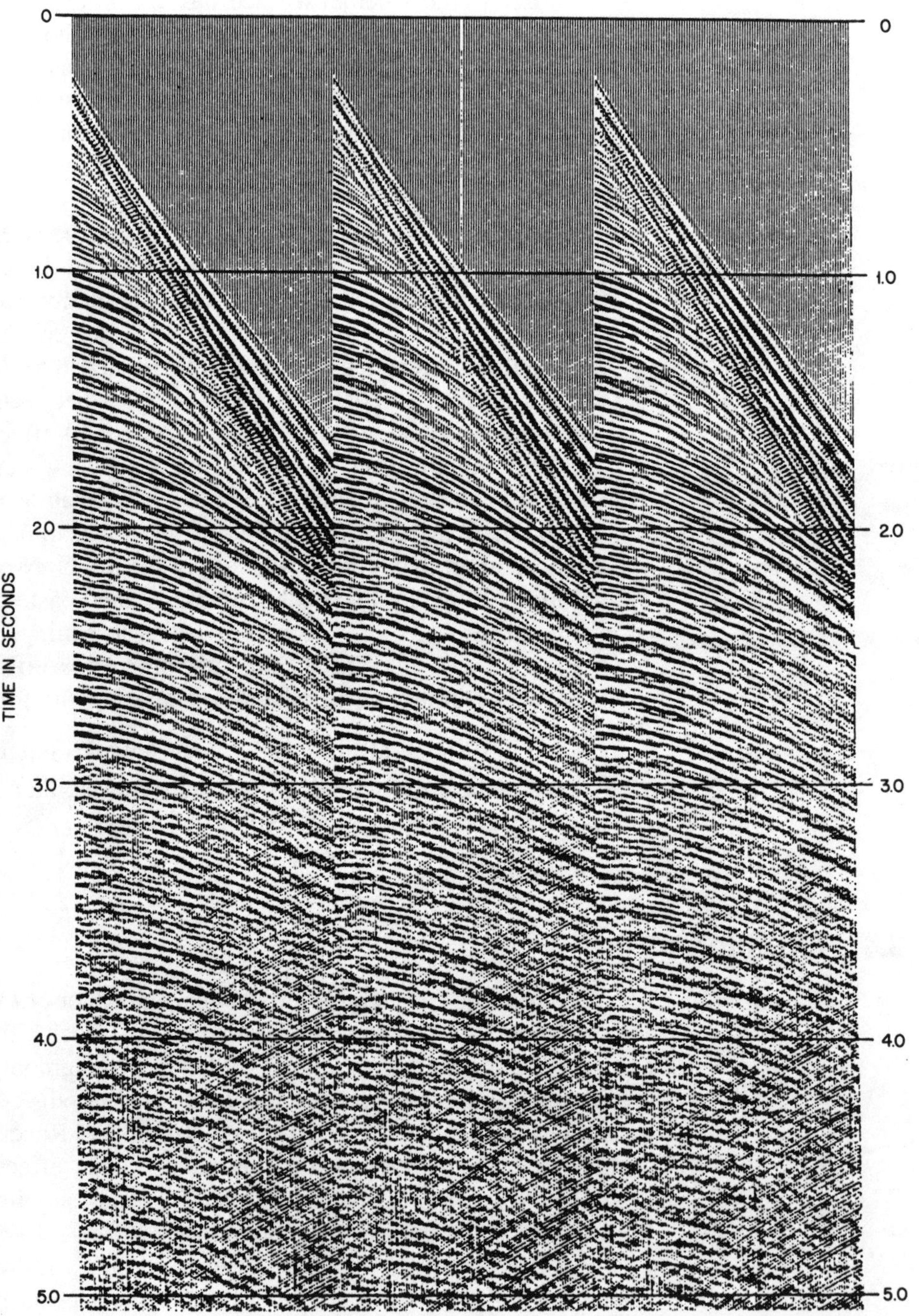

Fig. 5.1. Some shot records from the Dutch North Sea.

The water-bottom reflection crosses the dangler at a somewhat steeper angle (± 220 ms at shortest offset, ± 2170 ms at largest offset). Aliasing of this event owing to nonoverlapping receiver stations is clearly visible.

The result of (f,k_r) filtering of these shots with a passband of -4 ms to $+10$ ms per trace is shown in Figure 5.2. The first arrival and its dangler have been suppressed quite well and from underneath reflection hyperbolae have become clearly visible which illustrates that (f,k) filtering may increase the usable range of offsets. Without (f,k) filtering the first arrival and dangler if included in the range of offsets, would obscure the primary reflections up to 2 s (in this case).

The water-bottom reflection has not been removed properly by the (f,k) filter, illustrating that aliased energy may be hard to kill. Figure 5.3 shows that by using a narrower passband more aliased energy may be removed at the expense of the reflection hyperbolae uncovered in Figure 5.2.

Comparison of the three shots in Figure 5.1 shows that the first arrival and its dangler look very similar, indicating that the x_m dependence of this phenomenon is minimal. This similarity is caused by a horizontal sea-bottom and parallel bedding close to the surface. The first arrival and its dangler are examples of events in the three-dimensional wavefield that are (virtually) two-dimensional. Other examples of two-dimensional events are direct arrivals (cf. Section 3.8.1), reflections from a horizontally stratified earth, and water-bottom reverberations (in the case of a horizontal water-bottom). They are a function of x_o and just have a dc component in x_m.

Unwanted two-dimensional events with a constant apparent velocity such as the dangler in our example can be removed by one single narrow (f,k) filter, in (t,x_s), (t,x_r), (t,x_o) and even in (t,x_m).

In (t,x_s), (t,x_r) and (t,x_o) this single filter would be a narrow pie-slice around the apparent velocity of the event. The (f,k_m) filter that removes the linear x_o-dependent event in (t,x_m) is the filter which removes all energy at $k_m = 0$. This single filter would be just as effective as the other (f,k) filters in the other panels with respect to the linear event, but at the same time would be disastrous with respect to wanted horizontal line-ups.

If the trace spacing in the CMP were the same as in the CSP and CRP, then the same narrow velocity filter would remove the x_o-dependent event in either (t,x_r), (t,x_s) or (t,x_o). The trace distances are the same for center-spread shooting with shots half-way between receiver stations (Figure 4.4a). For off-end shooting, trace interpolation (dealiasing, Section 5.6), would have to be applied to arrive at the same trace distance.

The effect of this same (f,k) filter on other events would be quite different. The (f,k) filter in (t,x_s) or in (t,x_r) would have an x_m component, thus affecting the x_m component of other events, whereas the filter applied in (t,x_o) would not affect the x_m component of other events.

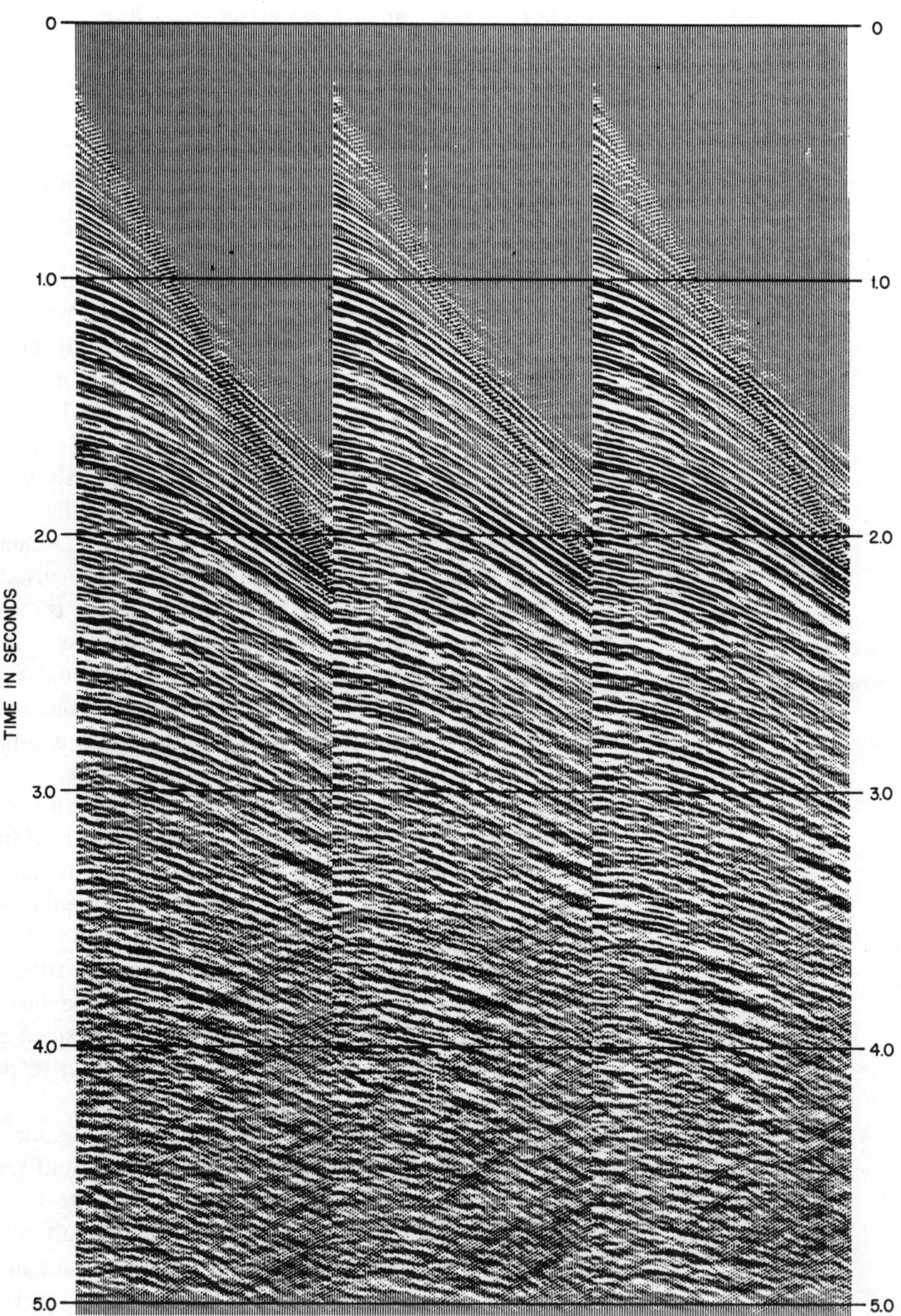

Fig. 5.2. (f,k_r)-filtered shots. Dips from -4 ms to $+10$ ms per trace have been passed. Note emergence of reflection hyperbolae that were still obscured by the first arrival and a dangler in Fig. 5.1. Note also aliased remnants of water-bottom reflection.

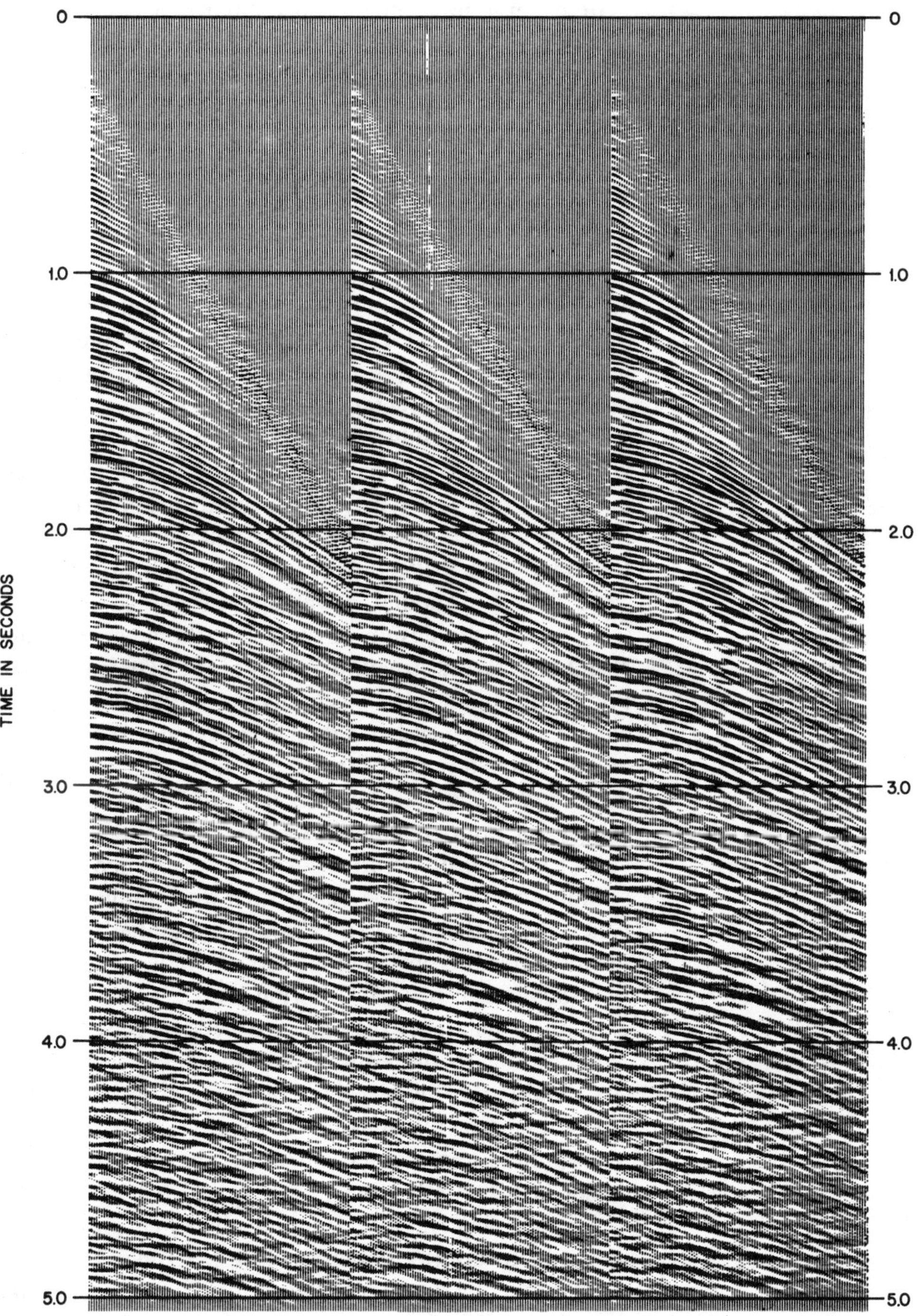

Fig. 5.3. (f,k_r)-filtered shots with a narrower passband than in Fig. 5.2. Dips from -2 ms to $+6$ ms per trace have been passed. Much desired signal energy has been removed.

The above argument shows that events that are a linear function of offset can best be removed in the CMP.

In general unwanted events (shot-generated noise) are not two-dimensional, so a three-dimensional filter $F(t,x_s,x_r)$ or $F_{tsr}(f,k_s,k_r)$ is usually required to tackle any unwanted part of the recorded wavefield. In practice, a three-dimensional filter is approximated by the application of two two-dimensional filters in succession, either in (t,x_r) and (t,x_s) or in (t,x_m) and (t,x_o), i.e., $F(t,x_s,x_r) \approx F_1(t,x_s) * F_2(t,x_r)$ or $F(t,x_s,x_r) \approx F_3(t,x_m) * F_4(t,x_o)$. Larner et al. (1983) demonstrate the effectiveness of the combination of (f,k_s) and (f,k_r) filtering for the removal of sidescattered energy. The cascaded application of a (t,x_r) filter and a (t,x_s) filter cannot be exactly equivalent to the cascaded application of a (t,x_o) filter and a (t,x_m) filter. The first filter will include all traces inside a square in (x_s,x_r) whereas the second filter will include all traces inside a square or rectangle in (x_m,x_o).

Good-quality results can be achieved by using cascaded filters in either way. One advantage of cascaded filtering in the midpoint/offset-coordinate system is some extra freedom in choosing the spatial length of the (f,k) filter in the COP. More filter points can produce a better performance of the filter, especially minimalization of smearing of the desired events. Another advantage of cascaded filtering in (t,x_o) and (t,x_m) is that the design of the (f,k) filters can exploit the separation of NMO effects and structure effects of the events to be passed by the filter. For "mild" geologies a velocity pass filter in (t,x_m) can be allowed to have a smaller passband than in (t,x_s) and (t,x_r), because the moveout of the reflection signal in $(t,x_{s,r})$ produces low apparent velocities in these cross-sections, whereas the mild geology creates only reflection energy for high apparent velocities in (t,x_m).

If (f,k) filtering before stack is merely aimed at improving the stacked section and if the same (f,k) filter were required in all common offset panels, then the (t,x_m) filter could be applied after stack with virtually identical results, provided the lateral velocity variation is not very large. A disadvantage of (f,k) filtering in (t,x_o) and (t,x_m) is the required de-aliasing of data shot off-end (see Section 5.6).

According to the equivalence theorem (see Section 5.11.5), (t,x_o), (t,x_s) and (t,x_r) filtering must be applied prior to NMO correction, otherwise one two-dimensional filter after stack could just as well be applied (unless improving the prestack data prior to the other prestack processes is essential).

Another way of approximating the three-dimensional filter $F(f,k_s,k_r)$ would be to Fourier transform $W(t,x_s,x_r)$ into $W_t(f,x_s,x_r)$ and then to apply a frequency dependent two-dimensional (x_s,x_r) filter. If the aim were to pass events in (t,x_s) and (t,x_r) with $V_{app} > V$, then the two-dimensional (x_s,x_r) filter could also be replaced by a high-cut x_s and a high-cut x_r filter with cut-off wavenumber $k_s = k_r = f/V$. Again, the same approach in (t,x_m,x_o) rather than in (t,x_s,x_r) would give somewhat more flexibility in selecting cut-off values.

Applying only a (f,k_r) filter to a data set obtained with off-end shooting will usually lead to an asymmetric result both in (t,x_m,x_o) and in the final stack. The asymmetry follows from similar arguments as used in Section 4.6.8 for asymmetric sampling. Hampson (1987) deals extensively with (f,k) filtering in two spatial domains, and the effect on symmetry. This very instructive paper also covers the effect of (f,k) filtering in various domains on the migration result.

For data recorded center spread, the application of only a receiver (f,k) filter leads to an odd/even effect in the CMP, in much the same way as discussed in Section 4.6.8 and shown in Figure 4.13b for asymmetric sampling.

5.5 Correction for Nonoverlapping Patterns

In Section 4.6.3 we noted that nonoverlapping patterns have their first zero-crossing at $k_{s,r} = 2k_N$ and not at $k_{s,r} = k_N$. Hence nonoverlapping patterns suppress energy only marginally around $k_{s,r} = k_N$, leading to aliasing of the recorded wavefield. Compensating for this aliasing may be considered at some point in the processing sequence. The simplest approach is to perform a two-trace running mix in both x_s and x_r. That this approach works can be seen from Figure 4.9. Adding the traces of four diamonds corresponding to two adjacent shots and two adjacent receivers is identical with the use of overlapping patterns.

Of course the mix should be a running mix, i.e., number of output traces is equal to number of input traces (apart from end effects). Halving the number of traces per spatial domain would double the sampling interval, hence halve k_N.

Figure 5.4 illustrates some aspects of correcting for nonoverlapping patterns for a situation in which the (shot or receiver) station spacing is three times the basic sampling interval, hence $k_N = k_{max}/3$. The nonoverlapping pattern has its first zero-crossing at $k_{s,r} = 2k_{max}/3$. Superimposed is the cosine response of a two-trace running mix with its first zero-crossing at $k_{s,r} = k_N$. The total pattern response is the product of the two-trace mixing filter and the response of the nonoverlapping field pattern. The upper part of Figure 5.4 shows wavenumber spectra of the continuous wavefield for $f = f_{max}$, $f = 2f_{max}/3$ and $f = f_{max}/3$. The energy left as a function of $k_{s,r}$ is found by multiplication of the (total) pattern response with the energy of the continuous wavefield. All energy above $k = k_N$ not suppressed by the total pattern response leaks (aliases) into $|k_{s,r}| < k_N$. As frequencies below $f = f_{max}/3$ carry no energy above $k = k_N$ (in this case) leakage does not occur for $f < f_{max}/3$.

It may be possible to do slightly better than the application of the cosine filter, though the greatest harm has already been done in the field. If there is much energy above $k_{s,r} = k_N$, but close to k_N, a longer filter may create a broader notch at $k_{s,r} = k_N$ than the cosine can. However, if significant energy is present for a wide range in $|k_{s,r}| > k_N$ a two trace-filter might be just as effective. Figure 5.4 also illustrates that the higher

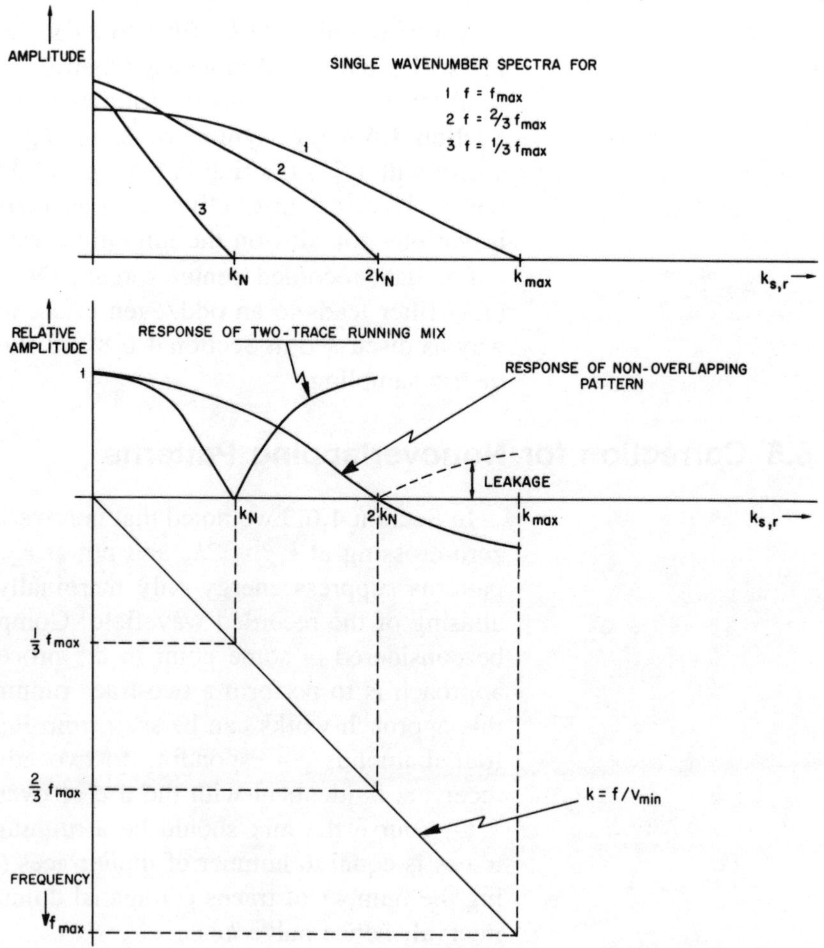

Fig. 5.4. Correction for nonoverlapping patterns. The lower part of the figure illustrates the effect of using a station spacing $dx_{s,r}$ which is three times as large as the basic sampling interval. A nonoverlapping pattern passes much energy above $k_{s,r} = k_N$ and also leaks energy above $k_{s,r} = 2k_N$. This is partly removed by a digital two-trace running mix. The top part of the figure illustrates some hypothetical wavenumber spectra of the continuous wavefield, prior to sampling, for various constant frequencies. There is no energy in these spectra for $k_{s,r} > f/V_{min}$. Hence, after sampling, leakage occurs only for frequencies $f > 1/3 f_{max}$.

frequencies are most vulnerable to the application of patterns. If the maximum usable frequency is much smaller than f_{max}, a digital high-cut frequency filter may also get rid of the aliased energy and reduce f_{max}.

As discussed in Section 4.6.6, patterns can have a detrimental effect on the desired signal. Hence the signal is better preserved if the correction for nonoverlapping patterns is carried out *after* NMO correction. Then, according to the equivalence theorem (Section 5.11.5), to perform the mixing after stack would save computer time with identical results.

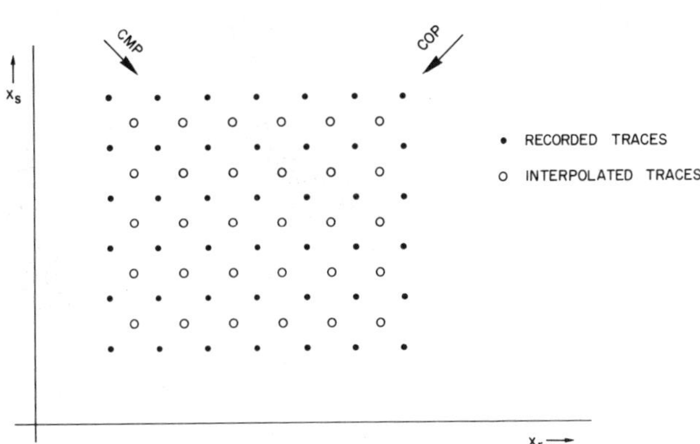

Fig. 5.5. Dealiasing of CMP and COP by horizontal interpolation. In the center of each square of recorded traces a new trace must be generated. This can be done by a two-dimensional horizontal interpolation filter.

On the other hand, successful (f,k) filtering requires unaliased or dealiased input, with the correction applied *before* NMO. The best solution will depend on the type of problem.

The correction for nonoverlapping patterns (if done pre-NMO) could be combined with dealiasing in (t,x_m) and (t,x_o).

5.6 Dealiasing of (t,x_m) and (t,x_o)

Subspaces (t,x_m) and (t,x_o) will be aliased in a regular off-end shooting geometry with $dx_s = dx_r \geq \Delta x_s = \Delta x_r$ (Section 4.5). This aliasing poses problems for prestack processes that have to be applied in the (t,x_m) and (t,x_o) subspaces [DMO, (f,k_m), (f,k_o) filtering]. Aliasing also leads to the checkerboard effect on stacked sections (Section 5.11.4). Hence dealiasing procedures are desirable.

If (t,x_s) and (t,x_r) have been sampled properly, these subspaces can be used to generate any trace within the (x_s,x_r)-recording window, using a two-dimensional time-invariant horizontal interpolation filter (Figure 5.5).

A much more convenient way of interpolation can be applied in (t,x_o) if the apparent velocities of all events are positive for $x_o > 0$, i.e., if the traveltimes of all events in the CMP increase with increasing offset. Then all energy for $x_o > 0$ maps into the right half of (f,k_o) (Figure 5.6a). After sampling, the aliased energy does not overlap the nonaliased energy (Figure 5.6b), which means that the data in (t,x_o) can be dealiased without recourse to (t,x_s) and (t,x_r). A simple dealiasing procedure would be to interpolate new x_o values along constant slowness lines $p_o = 1/2V_{min}$. In this direction the data are not aliased. Dealiasing in (t,x_o) leads automatically to dealiasing in (t,x_m).

Note that this procedure is another application of the generalized N-dimensional sampling theorem (Petersen and Middleton, 1962) as discussed in Section 4.5 and as also described in the first part of Bardan (1987).

Not all apparent velocities are always necessarily positive in the CMP for $x_o > 0$. Variations in the weathering velocity may lead easily to such differences in traveltimes of reflection events as to produce negative apparent velocities for small offsets. Especially deep reflections with

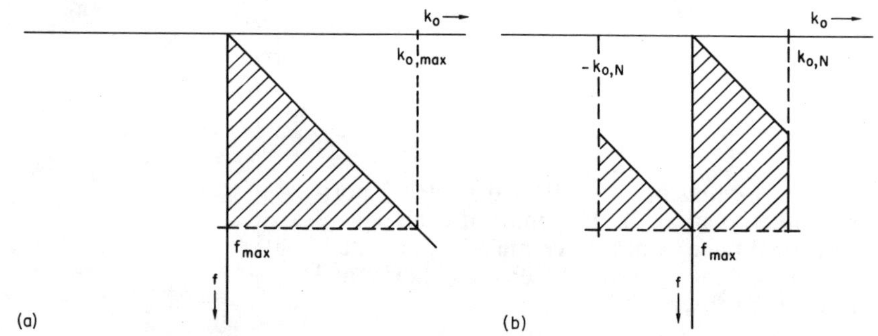

Fig. 5.6. Energy distribution in (f, k_o) for $x_o > 0$ if all events are dipping to the right in (t, x_o). (a) Continuous wavefield, (b) After sampling at the basic sampling interval in x_s and x_r aliased energy does not overlap nonaliased energy. This leads to a simple dealiasing procedure.

small NMO are prone to negative apparent velocities. In such cases there will also be energy in (f, k_o) for $k_o < 0$ and dealiasing can only be achieved properly by interpolation in the common-shot and common-receiver domains.

5.7 Statics

Variations in elevation and irregularities in the near-surface properties create irregularities in the recorded wavefield. The best solution to correct for the irregularities is to downward-continue the wavefield to a constant depth level below which the subsurface is more regular (Berryhill, 1979, discussed the principle for the zero-offset section). Prerequisites of a successful application of the redatuming technique are exact knowledge of the properties of the subsurface above the new datum and unaliased sampling of these rapid variations and the whole wavefield. Berryhill (1986) shows the successful example of wave-equation datuming applied to data recorded over a very irregular sea bottom formed by submarine canyons.

Usually the prerequisites for redatuming are not fulfilled. Then the solution is sought in applying surface-consistent static time-shifts to the data. Note that the application of static shifts corrects the wavefield by distorting it, because the application of static shifts is only an approximation to wave-equation redatuming.

The correction procedure for residual statics usually consists of the following steps: (1) measurement of relative time-shifts between traces, (2) determination of surface-consistent shot and receiver statics, and (3) application of static shifts to traces (Taner et al., 1974).

To avoid measuring shifts that are due to differences in moveout, NMO correction is normally applied prior to the static correction procedure. Relative time-shifts are measured in CMPs, CRPs, or CSPs. The COP is rarely used. Rather than measuring time-shifts in one space-time panel, consideration may also be given to measuring time-shifts of any trace with its eight nearest neighbors in all four space-time panels. This procedure leads to a redundant set of time-shifts for the whole recorded wavefield. The set of time-shifts has to be made consistent (which is easier said than done). Then all time-differences can be integrated to form a surface $t = t(x_s,x_r)$. At each point (x_s,x_r) t can be thought of as the sum of a shot static, a receiver static, a structural contribution, and a velocity bias, i.e.,

$$t(x_s,x_r) = t_s(x_s) + t_r(x_r) + t_m(x_m) + t_o(x_o). \qquad (5.1)$$

Each term in the right side of equation (5.1) transforms into a ridge along the corresponding axis in (k_s,k_r). Processing techniques can be applied in (k_s,k_r) to remove three ridges and to preserve one, thus obtaining the wavenumber spectrum of the shot statics and receiver statics. Around (0,0) the ridges meet each other, limiting the possibility of large-wavelength statics reconstruction. An approach similar to the one described here was given in Marcoux (1981).

5.8 Space-Conscious Deconvolution

Deconvolution is applied to sharpen reflection events or to remove reverberations and other multiple reflections. In present-day full-resolution or high-resolution recording the recorded noise such as groundroll may be many times stronger than the wanted signal, because the patterns are no longer optimal for groundroll suppression. Hence, if the deconvolution operator is determined on a trace-by-trace basis then the deconvolution process will mainly follow the spatial variations of the noise, which may have quite a different wavelet shape from that of the wanted signal. Instead, statistics should be gathered of the whole wavefield, so that spatial and temporal variations of the signal shape in (t,x_s,x_r) or in (t,x_m,x_o) can be separated from those of the noise. Significant progress is being made in this field, and many contractors now offer a surface consistent or channel-consistent deconvolution procedure based on an analysis of spatial variations of source signatures and receiver responses.

The shot signature in marine data acquisition using air guns is usually very stable. The sudden failure of an air gun may cause a discontinuity in the source signature. A space-conscious deconvolution technique will produce a stable signature and may identify the position where the gun failure occurred.

5.9 Velocity Determination

Stacking velocity determination and NMO correction together form a processing operation to which the three phases described in Section 5.1 apply.

The usual technique would involve: (1) measuring energy in the stack for a number of trial velocities along the seismic line, (2) analyzing results and establishing stacking velocity distribution $V_{\text{NMO}}(t,x_m)$ along the seismic line, and (3) applying NMO correction.

Most velocity determination techniques cannot be called space-conscious, as they usually investigate the energy in the stack only at a number of widely separated CMPs. Between these isolated points velocity distributions are established by linear interpolation. This procedure may easily lead to a loss of resolution. Space-conscious velocity determination will perform measurements at every CMP and then derive a smooth distribution that may vary quite rapidly in places, corresponding to any rapid spatial variations of the velocity distribution in the subsurface.

In more complex geologies with nonhyperbolic moveouts the term "stacking velocity" loses its meaning. Then the resolution of some events may not be good enough using conventional hyperbolic NMO correction and stack. In such situations it may be necessary to construct a velocity model for prestack migration.

An interesting alternative to the conventional hyperbolic NMO correction is presented in de Bazelaire (1988). Rather than searching for an optimum stacking velocity, this technique searches for the optimum time t_p as defined by the equation

$$(t + t_p - t_0)^2 = t_p^2 + x_o^2/V_1^2,$$

in which t is traveltime at x_o, t_0 is zero-offset traveltime, V_1 is a suitably chosen constant velocity, and t_p is the variable of the NMO analysis. This technique is computationally fast and would allow analysis at each CMP, hence making it a space-conscious process. It is too early to say whether this technique will gain acceptance in the industry.

5.10 NMO Correction

The aim of NMO correction is to line up primary reflections in order that after the correction their k_o wavenumber spectrum is zero except at $k_o = 0$. The apparent velocity V_o of the primary is made infinite. If the event was aliased in (t,x_o) prior to NMO correction, after NMO correction it will no longer be aliased.

In this section we analyze what effect the conventional hyperbolic NMO correction has on (1) frequency content of the NMO corrected trace, (2) apparent velocity V_o in the CMP, (3) wavenumber k_o, (4) apparent velocity V_m, and (5) wavenumber k_m.

The non-linear NMO correction is an operation in (t,x_o) that allocates the data of a seismic trace at time t to a new time t', according to

$$t' = \sqrt{t^2 - x_o^2/V_{\text{NMO}}^2}, \tag{5.2}$$

where the NMO velocity V_{NMO} (often called stacking velocity) is a function of t' and x_m:

$$V_{\text{NMO}} = V_{\text{NMO}}(t',x_m).$$

A window Δt of a trace at x_o is transformed by operation (5.2) into a new window $\Delta t'$ of the NMO-corrected trace. $\Delta t'/\Delta t$ represents the stretch factor S that can be derived by differentiating equation (5.2):

$$S = \frac{dt'}{dt} = \frac{t}{t'}\left(1 - \frac{x_o^2}{t'V_{\text{NMO}}^3}\frac{dV_{\text{NMO}}}{dt'}\right)^{-1}. \tag{5.3}$$

As $t' < t$ and V_{NMO} is usually an increasing function of t', equation (5.3) shows that the NMO correction indeed stretches the data ($S > 1$).

Note that a stretch of the time coordinate represents a contraction of the frequency axis, leading to the result that the spectrum of NMO-corrected traces has moved to lower frequencies f' with $f' < f$, according to

$$\frac{f}{f'} = \frac{dt'}{dt} = S. \tag{5.4}$$

Equation (5.4) can also be derived more formally by considering a pulse $g(t)$ at (t,x_o). The pulse is transformed by the NMO correction to a stretched pulse $g'(t')$ at (t',x_o) with

$$g'(t') = g(t/S). \tag{5.5}$$

Taking the Fourier transform of equation (5.5) and using a tilde to denote a Fourier transform, we find

$$\tilde{g}'(f') = S\tilde{g}(Sf), \tag{5.6}$$

confirming equation (5.4).

Equation (5.6) shows that the spectrum of the NMO-corrected trace is not only compressed, but also multiplied by the factor S. Dunkin and Levin (1973) use this result to demonstrate that NMO correction not only shifts the data to lower frequencies, but that it also increases the relative energy at lower frequencies, causing a relative loss of high frequencies in the stack.

In the following a constant V_{NMO} is assumed, neglecting its variation with t' and x_m. Then, using equation (5.3), equation (5.4) simplifies to

$$\frac{f}{f'} = \frac{t}{t'} = 1 + \frac{\delta t}{t'}, \tag{5.7}$$

where δt is the moveout at (t,x_o), $(t = t' + \delta t)$.

To investigate the effect of NMO correction on the apparent velocity V_o of an event $t = t(t,x_o)$, equation (5.2) differentiates with respect to x_o, bearing in mind that t is also a function of x_o.

$$\frac{dt'}{dx_o} = \frac{t}{t'}\frac{dt}{dx_o} - \frac{x_o/V_{\text{NMO}}^2}{t'}, \tag{5.8}$$

or　　　　　　　　　　　　　　　　　　　　　　　　　　　　　　(5.9)

$$\frac{1}{V_o'} = \frac{t}{t'}\left(\frac{1}{V_o} - \frac{x_o}{tV_{\text{NMO}}^2}\right). \tag{5.9}$$

This expression allows the computation of the apparent velocity V_o' of any event after NMO correction from its apparent velocity V_o before NMO correction.

Rewriting equation (5.2) as

$$t^2 = t'^2 + x_o^2/V_{\text{NMO}}^2, \tag{5.10}$$

results in a formula that describes hyperbolic events in (t,x_o) with parameters t' and V_{NMO}. The apparent velocity $V_{o_{\text{NMO}}}$ of the hyperbolic events can be found by differentiating equation (5.10) with respect to x_o:

$$\frac{1}{V_{o_{\mathrm{NMO}}}} = \frac{dt}{dx_o} = \frac{x_o}{tV_{\mathrm{NMO}}^2}.$$ (5.11)

Hence equation (5.9) can be rewritten as

$$\frac{1}{V_o'} = \frac{t}{t'}\left(\frac{1}{V_o} - \frac{1}{V_{o_{\mathrm{NMO}}}}\right).$$ (5.12)

Using equation (5.12) to compute the apparent velocity of the hyperbolic event after NMO correction indicates, not surprisingly, that this entity is infinite.

Introducing the *differential* apparent velocity V_{o_d} as

$$1/V_{o_d} = 1/V_o - 1/V_{o_{\mathrm{NMO}}},$$ (5.13)

i.e., $1/V_{o_d}$ is the difference in slowness of the event under consideration with the slowness $1/V_{o_{\mathrm{NMO}}}$ of the hyperbolic event, reduces equation (5.12) to

$$\frac{1}{V_o'} = \frac{t}{t'}\frac{1}{V_{o_d}} = \frac{S}{V_{o_d}}.$$ (5.14)

V_o' can also be considered as the differential apparent velocity *after* NMO correction, because it describes the difference between the flattened hyperbola and the NMO-corrected event. Equation (5.14) shows that the magnitude of the differential apparent velocity has increased by the multiplication factor S.

Using equation (3.19), we can now derive the relation between apparent wavenumber k_o' after NMO with apparent wavenumber k_o before NMO:

$$\frac{k_o'}{k_o} = \frac{f'}{f}\frac{V_o}{V_o'}.$$ (5.15)

Using equations (5.7), (5.13), and (5.14)

$$\frac{k_o'}{k_o} = \frac{t'}{t}\frac{V_o}{V_{o_d}}\frac{t}{t'} = \frac{V_o}{V_{o_d}}$$

$$= V_o(1/V_o - 1/V_{o_{\mathrm{NMO}}}) = 1 - V_o/V_{o_{\mathrm{NMO}}}.$$ (5.16)

Equation (5.16) describes the effect of NMO correction on wavenumber. For a hyperbolic event described by equation (5.10) $V_o = V_{o_{\mathrm{NMO}}}$, hence $k_o' = 0$, showing that all energy of such events is concentrated at $k_o' = 0$ after NMO correction. For steeply dipping events with $|V_o| \ll |V_{o_{\mathrm{NMO}}}|$, equation (5.16) shows that the wavenumber is hardly affected. Usually the wavenumber moves toward $k_o' = 0$, only if V_o has the opposite sign to $V_{o_{\mathrm{NMO}}}$ will the wavenumber of the event move outward. As argued in Section 5.6, events in (t,x_o) will usually have $V_o > 0$ for $x_o > 0$, unless there are large statics.

For a linear event with

$$t = x_o/V_o + \tau,$$

equation (5.16) is rewritten using equation (5.11), as

$$\frac{k_o'}{k_o} = 1 - \frac{V_o^2}{V_{\mathrm{NMO}}^2} \frac{x_o}{x_o + \tau V_o}. \tag{5.17}$$

For $\tau = 0$, an event passing through ($t = 0$, $x_o = 0$), equation (5.17) further reduces to

$$\frac{k_o'}{k_o} = 1 - \frac{V_o^2}{V_{\mathrm{NMO}}^2} = c. \tag{5.18}$$

This formula is independent of x_o and t, if V_{NMO} is constant, as we have assumed. Equation (5.18) is also derived in Ongkiehong and Askin (1988), and in Morse and Hildebrandt (1989). These authors use this result to investigate the effect of NMO correction on the response of the stack-array. This aspect is discussed in Section 5.11.6.1.

The effect of NMO correction on V_m can be derived by differentiating equation (5.2) with respect to x_m, bearing in mind that t is also a function of x_m:

$$\frac{dt'}{dx_m} = \frac{t}{t'} \frac{dt}{dx_m} = S \frac{dt}{dx_m} \tag{5.19}$$

(assuming again a constant V_{NMO}). Hence the apparent velocities in (t,x_m) are modified by the stretch factor S.

Using equations (3.19), (5.4), and (5.17) note that

$$\frac{k_m'}{k_m} = 1, \tag{5.20}$$

provided there is no lateral variation in V_{NMO}. For a laterally varying V_{NMO} k' would be modified depending on the derivative of V_{NMO} with respect to x_m.

Though the expressions for f', V_o', k_o', V_m', and k_m' as derived in this section look quite simple, all expressions do depend on the *local* variables x_o and t. Hence the expressions describe what happens locally to the frequency, apparent velocity, and wavenumber. For instance the contraction of k_o as described in equation (5.16) depends on the variable $V_{o_{\mathrm{NMO}}}$, a function that also depends on x_o and t [equation (5.11)]. The dependence of V_m on t' and x_m makes the effect even more complicated.

5.11 Stacking

5.11.1 Principle of the stacking method

A very important step in the seismic method is to obtain a zero-offset section of the wanted wave type (usually compressional waves) from which multiples, groundroll, converted waves, and other noise have been removed. This zero-offset section $Z(t,x_m)$ is a popular input to further processing, in particular to obtain a depth section.

The stacking method is the most commonly used technique for converting the recorded, preprocessed, and NMO-corrected three-dimensional wavefield into a two-dimensional section, which approaches the zero-offset section. The stacking method was introduced in Mayne (1962), as an alternative to long patterns for the suppression of surface waves.

Observe from Figure 4.2 that neighboring traces in the CSP are projected by stacking into neighboring positions in x_m. The same holds good for traces in the CRP. This means that an event that is coherent inside these panels will remain so in the stack. Normally an event that is coherent in one panel is also coherent in the other panels. An exception is ambient noise, which may be coherent in the CSP, but is not in CRP and COP. In the stack ambient noise will remain coherent if it was coherent in the CSP. Only at the edges of the CSPs as projected on the stack may the abrupt disappearance of an ambient noise event be expected. Tapering the x_o window toward the edges (i.e., weighting) will suppress the abrupt changes. This tapering will not improve the S/N in the stack unless S/N on the outer traces is lower than in the middle of the range.

5.11.2 The stack response

The stacking process can be described conceptually in much the same way as the action of patterns (Section 4.6.1) i.e., as a wavenumber filter:

(1) convolve each (t,x_o) panel with a filter. For a straight stack the filter coefficients are equal and the number of filter points is equal to the number of traces in the CMP, and

(2) keep only one output point of the convolution product for every CMP.

The second point describes the data reduction achieved by stacking, resampling the x_o axis to one point. The position of this output point is at the middle of the range of offsets [if the filter coefficients (the weights) are symmetrical with respect to the center of the range]. As a result, the character of the stacked trace bears more resemblance to the NMO-corrected traces in the middle of the offset range, than to the zero-offset trace, even though the reflection times in the stacked trace are close to those in the zero-offset trace.

The aim of stacking is to pass all energy of the desired primary reflection events (that have been NMO corrected) and to suppress everything else. Hence the desired response of the stacking filter is a spike at $k_o = 0$. This response is quite different from the desired response of a field pattern that has to pass all energy below $|k_{s,r}| = 1/2d_{s,r}$ and has to reject everything above that wavenumber. The scale of the patterns is also very different, and the filters are applied in different x domains. Furthermore, the stack response operates normally on the k_o spectrum as modified by the NMO correction [Section 5.10, equation (5.16)].

To what extent the stacking filter can be successful in rejecting everything above and below $k_o = 0$, can be analyzed using the stack response. The stack response can be considered as a z transform in x_o, with $z = \exp(2\pi j k_o dx_o)$ [dx_o is distance between traces in (t,x_o)]. Thus,

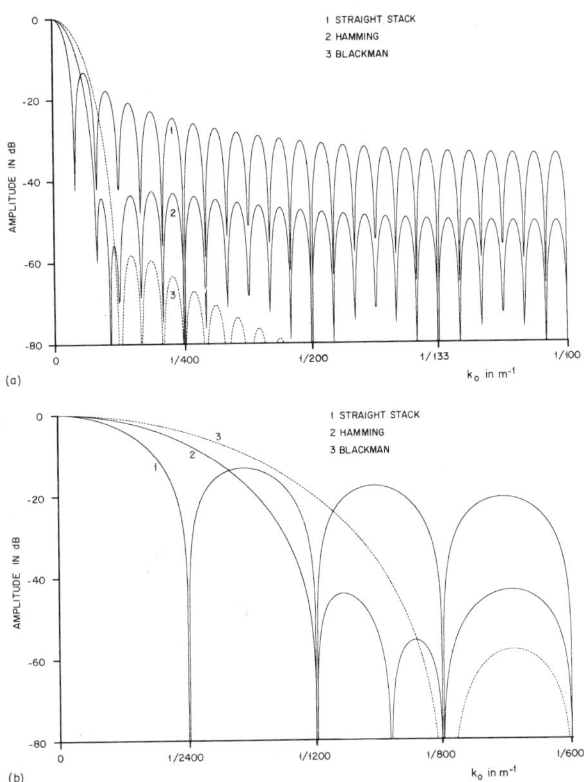

Fig. 5.7. Stack response in dB as a function of k_o for a shooting geometry with $|x_o|_{min} = 300$ m, $|x_o|_{max} = 2650$ m, $dx_o = 50$ m, and for different weighting functions. (a) Response to $k_o = 0.01$, (b) Expanded k_o-scale.

$$S(z) = \sum_{i=0}^{N-1} W_i z^i \bigg/ \sum_{i=0}^{N-1} W_i, \qquad (5.21)$$

N is number of traces in (t, x_o) and W_i is weight for trace i. Substitution of z in equation (5.21) leads to:

$$S(k_o) = \sum_{i=0}^{N-1} W_i \exp\left(2\pi j i k_o dx_o\right) \bigg/ \sum_{i=0}^{N-1} W_i. \qquad (5.22)$$

If all weights are equal, equation (5.22) can be simplified to a similar expression as equation (4.4) (omitting a lateral phase shift factor):

$$S(k_o) = \frac{\sin N\pi k_o dx_o}{N \sin \pi k_o dx_o}. \qquad (5.23)$$

In signal processing good practice usually is to taper toward the edges of windows used in processing. The filter coefficients or weights W_i can be used for this purpose, but they can also be designed to tackle particular noise types.

Figure 5.7a shows the stack response for a 48-fold stack with trace distance 50 m and first trace at 300 m. Three weight functions are used:

$$W_i = 1 \qquad \text{(straight stack)}$$

$$W_i = 0.54 + 0.46 \cos \theta_i \qquad \text{(Hamming)}$$

$$W_i = 0.42 + 0.5 \cos \theta_i + 0.08 \cos 2\theta_i \qquad \text{(Blackman)}$$

in which

$$\theta_i = 2\pi[x_{o_i} - (x_{o_{\max}} + x_{o_{\min}})/2]/(x_{o_{\max}} - x_{o_{\min}})$$

(more on weighting can be found in Oppenheim and Schafer, 1975, Ch. 5.5). Other approaches would leave a certain percentage of the weights flat and would taper only the extreme ends of the x_o window. In Figure 5.7b the same stack responses are displayed at an expanded scale close to $k_o = 0$. The various weights used are shown in Figure 5.8.

Note that the stack does not only pass energy at $k_o = 0$ but also close around it. Only for an infinitely long x_o window would the pass range be reduced to a spike at $k_o = 0$. The side lobes of the stack filter or leakage can be reduced to -40 or even to -80 dB at the expense of broadening the pass range.

[There seems to be a discrepancy between our conclusion that the stack does not only pass energy at $k = 0$ and a statement made by Fatti (1987) that "stacking can be shown to be equivalent to extracting the data along the $k = 0$ axis in the (f,k) plane." The latter statement applies to the *digital* wavenumber spectrum, whereas the stack response considers the effect of sampling and windowing on the originally *continuous* wavenumber spectrum. Windowing truncates each Fourier component resulting in a digital version with a dc ($k = 0$) component that is only zero if the wavelength of the Fourier component fits an integer number of times on the stack window. The notches ($-\infty$ dB) in Figure 5.7 represent wavelengths that fit exactly on the stack window.]

The benefit of tapering rapidly disappears if one or more traces are missing as is illustrated in Figure 5.9 which shows the stack response for the case with the trace at $x_o = 950$ m absent. This result illustrates

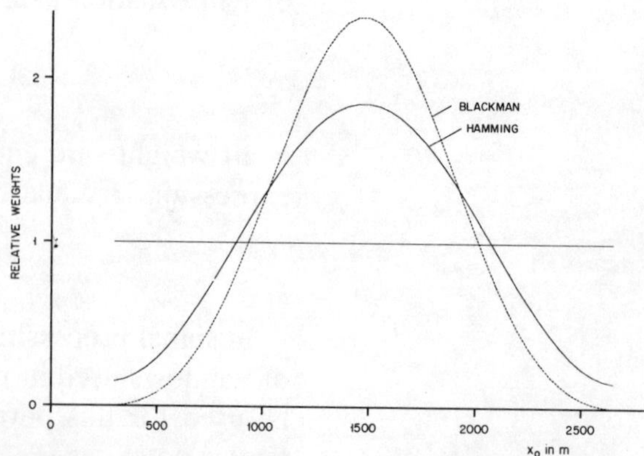

Fig. 5.8. Normalized weights corresponding to stack responses of Fig. 5.7.

the importance of the third phase of the editing and equalization process (Section 5.3).

For the suppression of random noise equal weights are best if the signal to random noise ratio is the same for all x_o. If this ratio varies and if it can be measured, optimum weighted stacking may be applied in which weights are proportional to a_i/N_i, where a_i is signal level in trace i, and N_i is energy of the noise in trace i. Optimum stacking methods have been described in Robinson (1970), White (1977) and Rietsch (1980). A crude but very effective weighting technique is the total blanking (muting) of the groundroll cone or first arrival.

With the improvements in data acquisition techniques, clearly much noise called "random" in the past was in fact aliased wavefield energy. Now this energy is often coherent and hardly aliased. Hence optimum weighted stacking will not be effective for modern data. It should still work for vertical stacking of Vibroseis signals as repeated recordings at the same surface position should be identical but for the ambient noise.

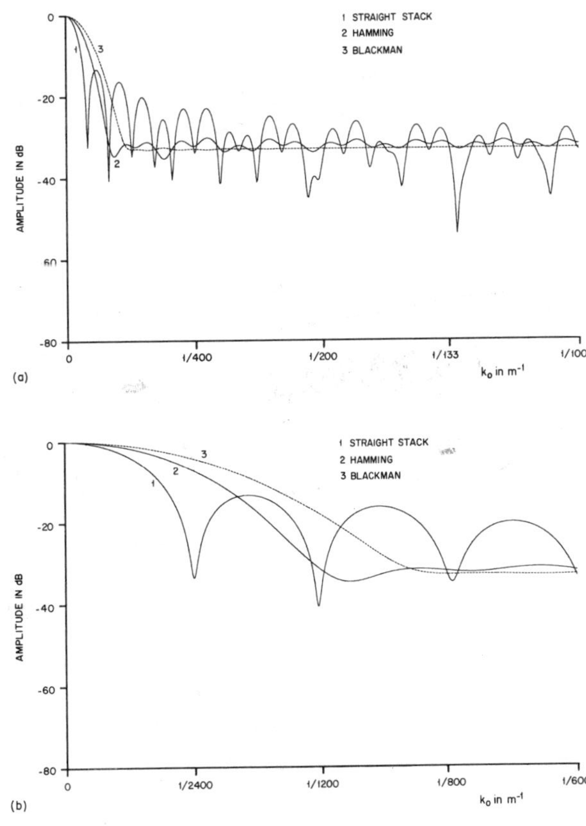

Fig. 5.9. Stack response in dB for same shooting geometry as in Fig. 5.7. Trace at $x_o = 950$ m is absent. (a) Response up to $k_o = 0.01$, (b) Expanded k_o-scale. Compare with Fig. 5.7 to see the influence of one missing trace.

5.11.3 Weighted stacking for multiple suppression

The stack response discussed in Section 5.11.2 describes the action of the stacking filter with respect to the k_o spectrum. To see the effect of stacking on multiples it is more convenient to compute the stack response as a function of the phase difference between primaries and multiples.

After NMO correction the residual moveout function of multiples is usually close to a parabola, i.e., proportional to x_o^2. A better suppression of the multiples is obtained by weighting the traces proportional to x_o rather than using equal weights. This quasi-linearizes the amplitude of the multiple as a function of x_o^2 and leads to a stack response similar to that for equal weighting of linear noise. A further flattening-out of the stack response can be obtained by using cosine tapering of the linear weights:

$$W_i = (0.54 + 0.46 \cos \theta_i)x_o \qquad \text{(Hamming)}$$

or

$$W_i = (0.42 + 0.5 \cos \theta_i + 0.08 \cos 2\theta_i)x_o \qquad \text{(Blackman)}$$

in which

$$\theta_i = 2\pi[x_{o_i}^2 - (x_{o_{max}}^2 + x_{o_{min}}^2)/2]/(x_{o_{max}}^2 - x_{o_{min}}^2)$$

In Figures 5.10a and 5.10b the various stack responses are plotted for a linear scale and a dB scale, respectively. Figure 5.11 shows the corresponding weights. Figure 5.10 shows that linear weights are always better than equal weights and that tapering pushes the ripples down at the expense of broadening the passband. As random noise (if equally distributed over x_o) is best suppressed by equal weights, tests may be required to find the best compromise between random noise suppression and multiple suppression.

Figure 5.12 shows a comparison of the same section with and without the use of linear weights. The strong reflection bands around 0.9 and 1.6 s are followed in Figure 5.12a by strong multiples in the region of 2.35 s and deeper. In Figure 5.12b the indicated reflection at 2.7 s is better resolved than in Figure 5.12a, whereas the zone above and below this reflection has been cleaned up.

5.11.4 The checkerboard effect

The checkerboard effect is a product of regularly differing offset combinations in neighboring CMPs. For off-end shooting with equal shot and receiver intervals every other trace of the stacked section looks different. For off-end shooting with shot intervals twice as long as the receiver intervals, the effect appears in groups of four, etc. Checkerboarding in pairs is also called the odd/even effect.

The stack response of two neighboring CMPs differs by a factor of $\exp(-2\pi jkdx_o)$, (a shift in x_o of dx_o, $dx_o = dx_r$) which for events with a linear time-distance relationship in (t, x_o) leads to a time shift $\Delta t =$

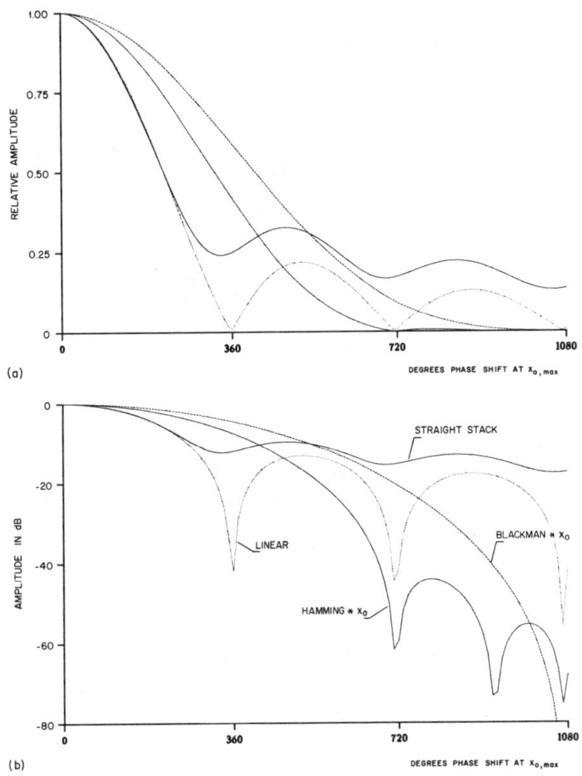

Fig. 5.10. Stack response as a function of phase difference between $x_o = 0$ and $x_o = 2650$ m for a shooting geometry with $|x_o|_{min} = 300$ m, $|x_o|_{max} = 2650$ m, $dx_o = 50$ m, and for different weighting functions. (a) linear amplitude scale, (b) dB scale.

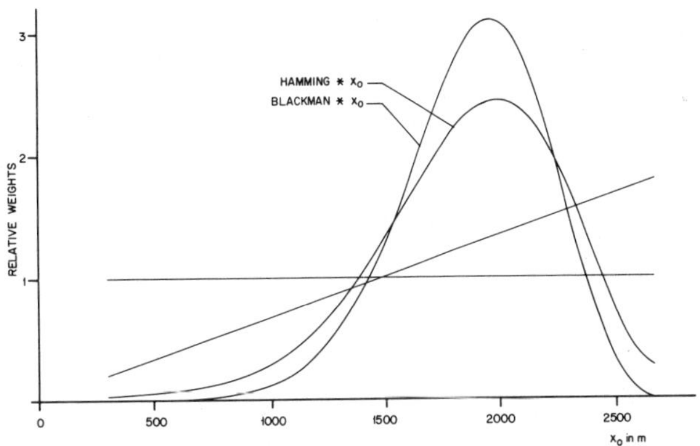

Fig. 5.11. Normalized weights corresponding to stack responses of Fig. 5.10.

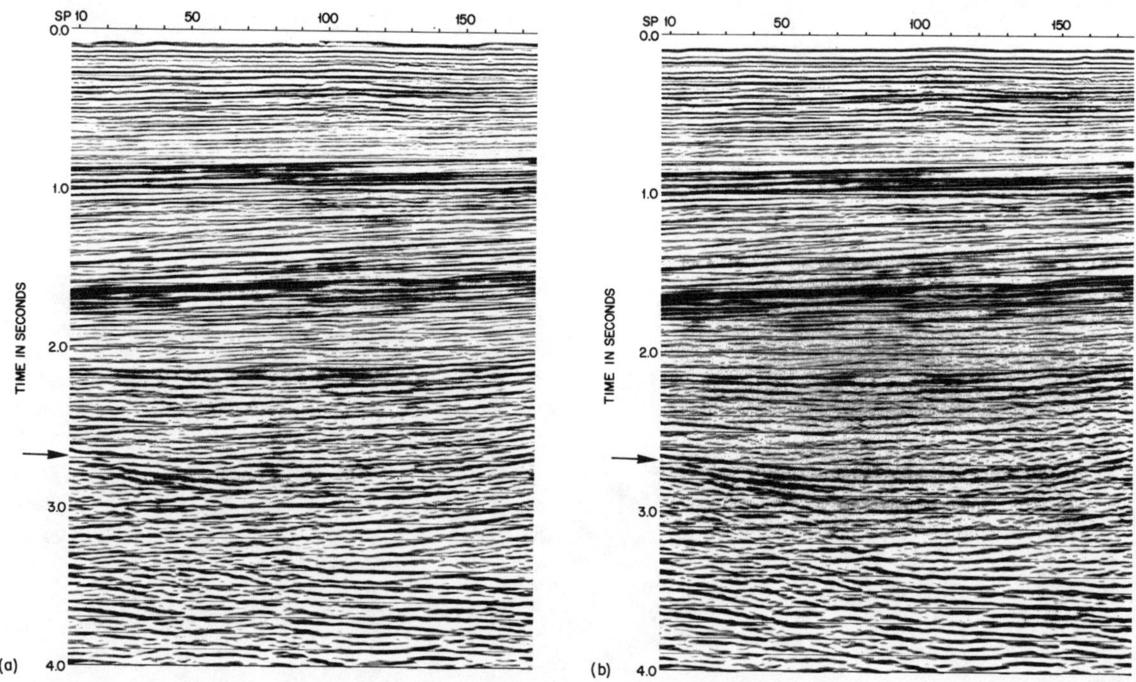

Fig. 5.12. Comparison of stacks. (a) No weights, (b) Weights proportional to x_o.

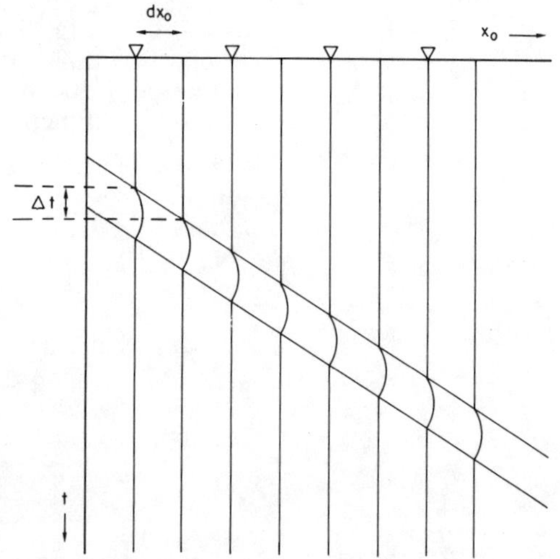

Fig. 5.13. For an event with constant apparent velocity V_o separate stacking of odd- and even-numbered traces into neighboring CMPs leads to a checkerboarding pattern with $\Delta t = dx_o/V_o$.

dx_o/V_o between neighboring CMPs (Figure 5.13). In particular first arrivals, as shown in Figure 5.1, and multiples or primaries with residual moveout may create a checkerboard-like pattern in the stack at shallow levels. Thus the checkerboard pattern also has some diagnostic value. At deeper levels it may be observed as a regular variation in amplitudes. A satisfactory remedy is to dealias the CMP prior to stacking (Section 5.6). Cosine tapering of the x_o window, as discussed in Section 5.11.2, also reduces the effect.

After stack the odd/even effect can be removed by application of a lateral highcut filter in (t,x_m). A crude solution is to binate the stack. In that case an antialias x_m filter has to be applied first. As checkerboard noise does not fit the migration model, migration will also reduce the effect.

5.11.5 The equivalence theorem

Processing geophysicists often need to explain to their clients the relative merits of wavenumber filtering, frequency filtering, and (f,k) filtering before or after stack. The explanation can be given using the equivalence theorem, which may be stated as:

The output of stacking immediately followed by the application of a (t,x) filter is identical with the output of the application of a (t,x) filter with exactly the same filter points in CSP, CRP, or COP, immediately followed by stacking, provided:

(1) the (t,x) filter does not vary, and

(2) the filter operation is applied after "missing" traces, missing shots, and missing receivers have been included in the prestack data set as traces with zero amplitude.

Missing traces include trace positions outside beginning and end of shot and receiver panels up to the half-length of the x component of the (t,x) filter. For regular off-end shooting consecutive traces of each common offset panel project onto even or odd positions in the stack. Then the equivalence theorem applies if we define a missing trace half-way between each pair of traces in the COP.

Bearing in mind that CSP traces and CRP traces project to neighboring positions in the stack, the derivation of the equivalence theorem is straightforward.

Expressed in a formula, the equivalence theorem reads:

$$\left[\sum_{x_o} U(t,x_m,x_o) \right] * g(t,x) = \sum_{x_o} [U(t,x_m,x_o) * g(t,x_i)], \quad i = s,r,m, \quad (5.24)$$

in which $U(t,x_m,x_o)$ is the NMO-corrected prestack wavefield and $g(t,x)$ is the (t,x) filter.

The equivalence theorem can be extended also to include (t,x) filtering in the CMP, which is equivalent to frequency filtering after stack in which the frequency filter is the "stack" of the (t,x) filter. This frequency filter would reduce to one point for x filtering (without time dependence), hence x filtering in the CMP (after NMO) followed by straight stacking does not change the stack (as opposed to weighting the traces in the CMP).

The application of a (t,x) filter in the CSP followed by another (t,x) filter in the CRP is equivalent to the application of the convolved filter after stack.

I have been using the term (t,x) filtering rather than (f,k) filtering to indicate convolution in (t,x), rather than multiplication in (f,k). With (f,k) filtering it is more difficult to make end effects the same before and after stack. The equivalence theorem would still be valid, if *identical* were replaced by *very similar*. Examples of this application of the equivalence theorem were presented in Engstrøm and Hvidsten (1986).

If NMO correction is applied after (t,x) filtering before stack, the equivalence theorem no longer applies, as the NMO correction is a nonlinear operation.

Usually the number of traces contributing to the stack is time-variant owing to initial blanking. Stacking programs take care of this by applying a time-variant scaling factor. Hence the equivalence theorem does not apply for the t component of the filter $g(t,x)$ in the shallow part of the data.

The equivalence theorem tells us that for deep data with only small moveout, 2-D filtering after stack can produce a result comparable to 2-D filtering before stack.

The equivalence of (t,x_m) filtering in the COP with (t,x_m) filtering in the stack using the same filter throughout is very easy to see. This equivalence is still valid for the x_m component of the filter if the NMO correction is applied between filtering and stacking and if the NMO correction is independent of x_m [for a constant-velocity distribution equation (5.16) applies]. It is not valid for the t component of the filter because of the stretching effect of the NMO correction. However, for small stretch factors the deviations are small. Therefore, the most economical way of producing an (f,k) filtered stack is:

(1) (f,k_o) filter (including dealiasing in the case of off-end shooting, see Section 5.6),

(2) NMO correction and stack, and

(3) (f,k_m) filter.

This procedure will produce an end result that is virtually always similar to or better than the cascaded application of (f,k) filters in CSP and CRP before NMO and stack.

However, it may be necessary to improve the quality of the prestack data to allow better determination of velocity, statics, and deconvolution parameters. Then (f,k) filtering in (t,x_o) *and* (t,x_m) should be considered.

5.11.6 Stacking and the field technique

5.11.6.1 The total response of field patterns and stack

If the seismic wavefield has been recorded using the basic sampling interval for both shots and receivers, then the stack response (Section 5.11.2) fully describes the effect of stacking on the NMO-corrected wavefield. The stack response is particularly useful in describing the effect on groundroll with its wide range of $k_{s,r}$ values.

However, normally much larger than basic sampling intervals are used, with shot and receiver patterns laid out to carry out a filtering and re-sampling operation prior to recording (Section 4.6). Investigating the combined effect (total response) of field patterns and stack on the continuous wavefield $W(t,x_s,x_r)$ is of interest.

Section 4.6.2 has shown that the combined effect of a shot pattern and a receiver pattern is a two-dimensional pattern response in (k_s,k_r) and in (k_m,k_o). The stack response discussed in Section 5.11.2 is a function of k_o only. If the NMO correction is neglected for the time being, then the total response S_{tot} can be written as

$$S_{tot}(k_m, k_o) = p_s \times p_r \times S(k_o), \qquad (5.25)$$

in which $p_s \times p_r$ follows from equation (4.5b) and $S(k_o)$ from equation (5.23). Note that the pattern length of $S(k_o)$ is much larger than the length of the field patterns.

The field pattern part of equation (5.25) is illustrated in Figure 5.14a for the situation of shot and receiver pattern both having 25 elements spaced at 2 m. Only the first side lobes of shot and receiver patterns are included fully. Figure 5.14a is to be compared with Figure 4.7b. The shot and receiver patterns have zero-crossings $(-\infty$ dB) along $k_s = 1/50$ m^{-1} and $k_r = 1/50$ m^{-1}, producing a diamond shape in (k_m,k_o). Each color in Figure 5.14a and in the following Figures 5.14b thru f represents a range of 6 dB.

The pattern response shown in Figure 5.14a represents the effect on the continuous wavefield of using 50 m length spatial filters, *prior to resampling*. If a shot and receiver sampling interval (station spacing) is used of 25 m, the first zero-crossings of shot and receiver patterns coincide with k_{s_N} and k_{r_N}, respectively. All remaining energy outside these wavenumbers folds back into the central passband.

A discussion of the total response for cases 1 through 5 follows

Case	Shot Pattern	Receiver Pattern	CMP Pattern	NMO	Figure
1	25 × 2	25 × 2	24 × 50	no	5.14b
2	no	25 × 2	24 × 50	no	5.14c
3	25 × 2	25 × 2	12 × 100	no	5.14d
4	no	25 × 2	12 × 100	no	5.14e
5	25 × 2	25 × 2	24 × 50	yes	5.14f

Case 1. Application of the stack response to Figure 5.14a is illustrated in Figure 5.14b for 24 traces spaced at 50 m. The stack response produces zero-crossings parallel to $k_o = 0$. The first alias (secondary passband) occurs at $k_o = \pm 1/50$ m^{-1}. Along these lines k_o energy suppression is taken care of by the field patterns only. Inside the passband of the field patterns (the diamond in Figure 4.7b) the suppression is mostly taken care of by the stack response. In this area the suppression is better

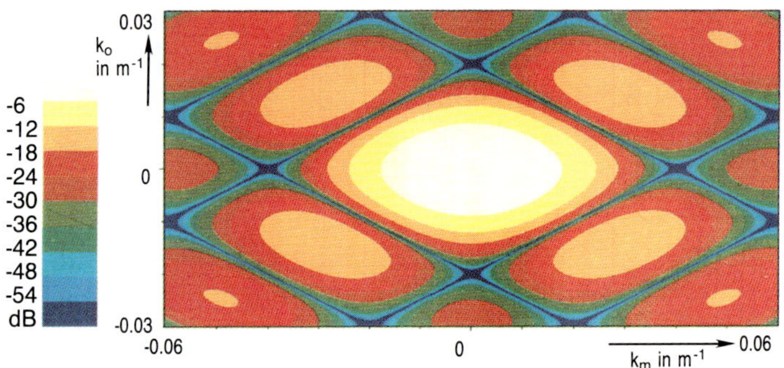

Fig. 5.14a. Response of the combination of shot patterns and receiver patterns each having 25 elements spaced at 2 m.

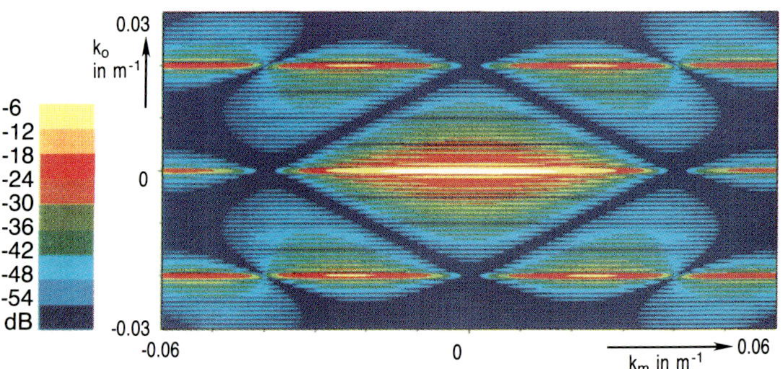

Fig. 5.14b. Combined response of shot and receiver patterns as in Fig. 5.14a, and a CMP pattern of 24 traces spaced at 50 m.

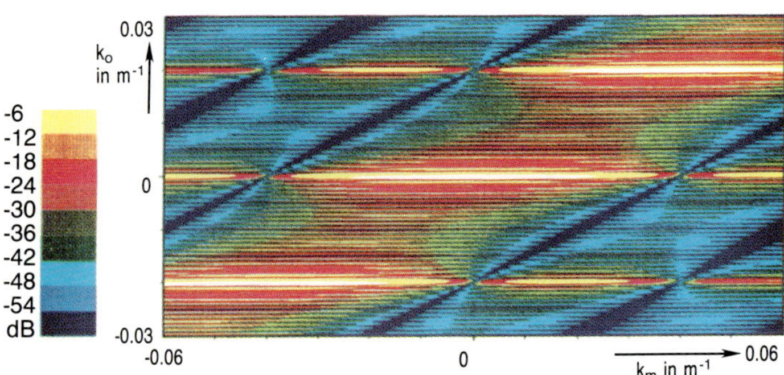

Fig. 5.14c. As Fig. 5.14b, but without a shot pattern.

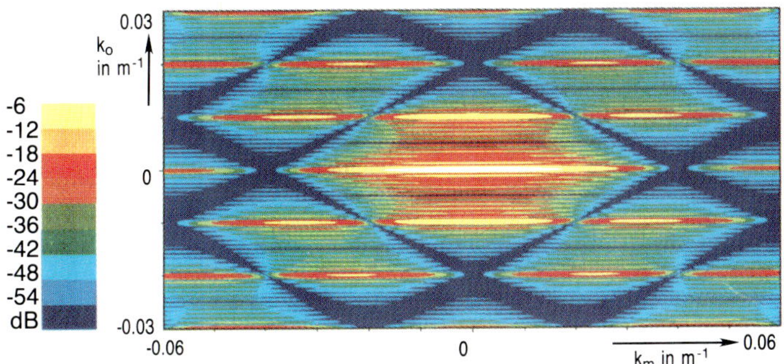

Fig. 5.14d. Combined response of shot and receiver patterns as in Fig. 5.14a, and a CMP pattern of 12 traces spaced at 100 m.

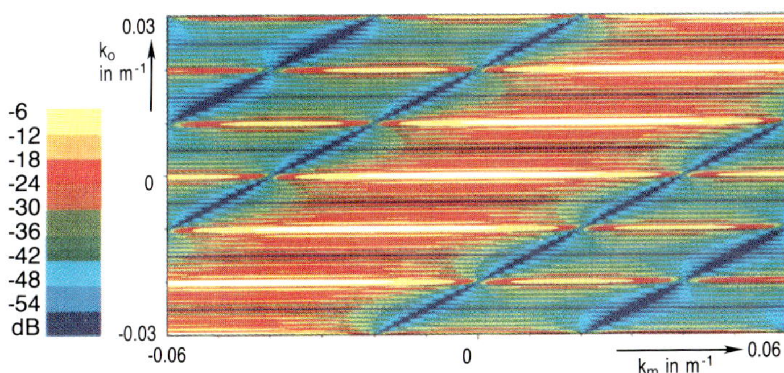

Fig. 5.14e. As Fig. 5.14d, but without a shot pattern.

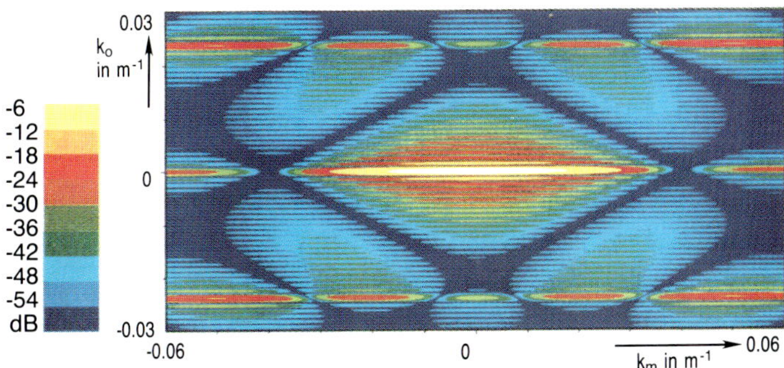

Fig. 5.14f. Combined response as in Fig. 5.14b, modified for the effect of NMO correction with velocity 1500 m/s on a linear event with apparent velocity $V_o = 600$ m/s.

than 24 dB for $|k_o| > 3/1200$ m^{-1}. How much energy of $W(t,x_s,x_r)$ is passed by this filter depends of course on the double wavenumber spectrum $W_{sr}(t,k_s,k_r)$. In the hypothetical case of a horizontal earth with no inhomogeneities, all energy is concentrated along the $k_m = 0$ axis (Figure 3.11). The zero-crossings of the field pattern then nicely coincide with the energy passed by the alias bands of the stack response. However, in realistic cases there is also much energy in other parts of the (k_m,k_o) plane. In particular, backscattered and sidescattered groundroll has its energy spread over large k_m values.

The best suppression of energy is in areas of (k_m,k_o) where all three patterns have low responses. Such an area extends over a broad range along the k_o axis, where suppression is greater than 54 dB. Figure 5.14b shows the multiplication of the straight-stack response with the field pattern response. Selecting a weighted stack, such as shown in Figure 5.7 would produce much better suppression over a large range in (k_m,k_o) at the expense of broadening the passband and the alias bands.

The worst suppression of energy is in areas where the passband of a field pattern (along a diagonal in Figure 5.14b) coincides with an alias band of the stack pattern. There suppression is locally less than 18 dB. Figure 4.8a shows the passbands of the field patterns over a wider range. These secondary passbands are responsible for groundroll, that may even appear in data recorded with both shot and geophone patterns.

Figure 5.14b has been computed for a rather small multiplicity and large shot and receiver intervals. (Larger multiplicity would have required a larger picture to represent all the loops of the stack response). If smaller intervals are chosen, then the scales of Figure 5.14b change, but the general shape remains the same. The effect of smaller intervals is to pass more energy of signal and shot-generated noise, as the first zero-crossings of the field patterns move outward. As a result the residual shot-generated noise in the straight stack will in general be stronger for short shot and receiver intervals than for large intervals. The stack pattern on its own is not as effective as the action of the two field patterns. Hence with finer spatial sampling it becomes more important to apply (prestack) spatial filters in addition to stacking for optimum suppression of the groundroll or other noise.

Case 2. Figure 5.14c illustrates the effect on the total response of not using a shot pattern. A broad band is shown in which there is very little suppression leading to severe aliasing after sampling.

Case 3. Off-end shooting with equal shot and receiver station spacing leads to CMPs in which the traces have a distance double that of the shot and receiver interval. If no precautions are taken to remedy this situation (Section 5.6), then the stack response has an alias band inside the pass area of the field patterns. Figure 5.14d illustrates the effect for a 12 point CMP pattern with trace spacing of 100 m and also shows that this configuration leads to considerably more aliased energy than is the case for 24 traces at 50 m.

Case 4. Figure 5.14e illustrates the case for no shot pattern and 12 traces at 100 m. Note that much energy may be left untouched in this case leading to aliased data after sampling.

Case 5. The total responses shown in Figures 5.14b thru e correspond to the (τ, $p_o = 0$) stack, i.e., a stack without NMO correction, also called plane wave stack. [Claerbout (1985) discusses this stack. Taner (1976) introduced the plane wave common shot stack, or (τ, $p_r = 0$) stack.] Normally, an NMO correction will be applied prior to stacking. As discussed in Section 5.10, the NMO correction has a complicated effect on the various events in the wavefield. Usually NMO correction moves data points toward the k_m axis. Only for linear events passing through ($t = 0$, $x_o = 0$) will there be a uniform contraction of the wavefield along the k_o axis, according to equation (5.18). This contraction is followed by stacking, hence, *relative* to the original wavefield, the stack response moves toward larger wavenumbers. Figure 5.14f illustrates the effect for the case of a linear event with apparent velocity $V_o = 600$ m/s and NMO velocity of 1500 m/s [$c = 0.84$, equation (5.18)]. Comparing this result with Figure 5.14b shows that the first notches of the field patterns no longer coincide with the first alias of the stack response at the $k_m = 0$ axis. Yet, the suppression of energy falling along the k_o axis is still quite large (>30 dB) for all larger k_o values. Close to $k_o = 0$ (large wavelengths) there is less suppression due to the relative broadening of the main passband of the stack response. Further away from the k_o axis the effect of NMO correction is to move the alias band of the stack response to larger $|k_o|$ ranges, basically moving areas of less good suppression to other parts of the (k_m,k_o) plane. Seemingly the net effect of NMO correction is a minor overall deterioration in the suppression of low velocity events. How serious the deterioration actually is depends on the energy distribution in the continuous wavefield (is most of the groundroll energy concentrated around the k_o axis?).

5.11.6.2 The stack-array approach

In 1986 Anstey's important article on the *stack-array approach* was published in *The Leading Edge* and was followed by a lively discussion (see Anstey, 1986a, 1986b, 1986c, 1987, and 1989). Morse and Hildebrandt (1989) discuss in detail the effect of NMO on the stack-array performance.

According to Anstey (1986c) the stack-array approach can be summarized as follows:

"(1) The stack-array criterion: an even, continuous, uniform succession of geophones across the CMP gather,

(2) Marine operation, end-on spread: group length equal to group interval, source interval half the group interval.

(3) Land operation, split spread: group length equal to group interval, source interval equal to group interval, shot between the groups."

The stack-array approach clearly offers great improvements over earlier recording techniques, in which, for instance, the source interval was four times as large as the receiver interval. Morse and Hildebrandt (1989) demonstrate the advantage of an even geophone distribution over the CMP as compared to an uneven distribution.

As discussed in Section 4.11.1, the shot station interval in marine data acquisition cannot be made arbitrarily small due to limitations in the ship's speed. The hydrophone patterns can usually be made shorter than the shot interval, hence the shot station interval is the limiting factor in the spatial resolution of marine seismic data. For a given shot station interval the stack-array concept prescribes a twice as long group length. This recommendation leads to the configuration described in Figure 4.3b. At closer inspection the receiver station interval is also half the group length in this configuration. Taking as an example a 30 m shot interval, the group length is 60 m and the distance between traces in CSP, CRP, and CMP is 60 m, whereas the distance between traces in the COPs (= CMP interval) is 30 m. However, with the same shot interval and the same acquisition costs a higher spatial resolution could be achieved by splitting up the receivers in 30 m groups, leading to a CMP interval of 15 m with the same multiplicity as before. The N-dimensional sampling and interpolation theory (Sections 4.5 and 5.6) asserts that interpolation will then allow uniform sampling of COPs (15 m interval between traces) and CMPs (30 m interval between traces) thus producing the best resolution for given acquisition costs.

The importance of the shot pattern has been generally overlooked in the stack-array discussion. The three-dimensional nature of the continuous wavefield, and in particular of the groundroll (cf. Figures 1.1–1.4), was not taken into account, leading to criterion (1) of the stack-array approach that does not specify anything about shot patterns. Section 5.11.6.3 describes a better alternative.

5.11.6.3 The optimum field technique

Section 5.11.6.1 discussed the dramatic difference between Figures 5.14b and c, and also between Figures 5.14d and e, determined by the absence or presence of a shot pattern. Hence, by using the *symmetry criterion* (see also Section 4.6.8) better data than possible with the stack-array approach can be recorded.

Equal shot and receiver intervals,

Equal length shot and receiver patterns, and

Equal number of elements in shot and receiver patterns.

This criterion applies to both marine and land data acquisition. For center-spread recording on land, it would only be necessary to shoot into every other station (Figure 4.4b), though it is far more convenient to record all receiver stations. The "hands-off acquisition technique" described in Ongkiehong and Askin (1988) uses basically the same criterion.

Symmetry between shot sampling and receiver sampling ensures that there is not only an even and uniform distribution of geophones over the seismic line (stack-array approach), but also an even and uniform distribution of shots. Symmetric sampling prevents asymmetry of diffractions in the stack and also prevents asymmetry in the CMPs of center-spread data (Section 4.6.8).

Symmetry in itself is not a guarantee for good quality, the spatial sampling intervals should also be sufficiently small for an alias-free recording of the desired signal. Hence adherence to the requirements of full-resolution recording (Section 4.9) gives the best chance for good quality data. This technique can now be summarized as:

(1) Shot and receiver intervals equal to basic signal sampling interval,
(2) If basic sampling interval smaller than basic signal sampling interval use shot and receiver patterns according to the symmetry criterion.

(For the definition of basic sampling interval and basic signal sampling interval see Section 4.2.) Microspreads can be usefully applied to determine the basic sampling interval for the area of interest (Section 4.10.1).

Note that the basic sampling intervals are inversely proportional to the maximum frequency of interest. Though actual recording may take place at 2 ms with Nyquist frequency of 250 Hz, the required or achievable frequency may be much lower, perhaps 70 Hz, thus considerably reducing the required spatial sampling effort.

5.11.7 Toward a better zero-offset section

The normal stacking process has a number of disadvantages:

—complex overburden causes nonhyperbolic moveout,
—the NMO correction distorts the frequency spectrum of the wavefield,
—diffraction energy is suppressed and crossing events usually have different stacking velocities, and
—the stacked section is not a true zero-offset section.

For these reasons many researchers have been, and still are looking for techniques that allow the creation of a better approximation to the zero-offset section.

One of the alternative techniques is slant-stacking or plane wave stacking. Taner (1976) considered plane wave stacks of the CSP, i.e., (τ, p_r) transforms with $p_r = C$, repeated for all CSPs. Slant CSP stacks with $C \neq 0$ enhance dipping events relative to horizontal events (Sheriff and Geldart, 1983).

Claerbout (1985) discusses the plane wave stack for the CMP with $p_o = 0$, i.e., CMP stacking with infinite stacking velocity, and argues that this stack is very close to a zero-offset section, and therefore can be migrated as if it were.

More elaborate uses of slant stacking involve first a full (τ, p_o) transform of each CMP, followed by stacking along the elliptical moveout (EMO) in the (τ, p_o) plane. Kelamis and Mitchell (1989) and Mitchell and Kelamis (1990) give examples of the application of a technique that also involves a hyperbolic velocity filtering step in the transformation to (τ, p_o). According to Kelamis and Mitchell this technique produces a better approximation to a zero-offset section than NMO stacking.

The dip moveout (DMO) correction procedure is another technique to create a better zero-offset section by equalization of common offset panels prior to stacking. This procedure acknowledges that stacking without DMO correction is not a common reflection point technique (as it was originally called) but a common midpoint technique, i.e., the reflection point at zero-offset differs from that at other offsets in case of a non-horizontally layered earth. The paper by Judson et al., 1978, described the Devilish procedure to correct for the attenuation of dipping reflections and diffractions. Later authors (Yilmaz and Claerbout, 1980, Deregowski and Rocca, 1981, Bolondi et al., 1982) described other DMO techniques. See Hale (1984) for a survey of DMO techniques and a description of a Fourier transform approach to DMO. Berkhout (1986) predicts that "wide-angle CRP stacking" (an alternative expression for DMO) will "largely replace the conventional CMP stack."

A further discussion of these improvements is beyond the scope of this text. Claerbout (1985) is recommended for further reading. Claerbout's book builds on the three-dimensional nature of the single 2-D seismic line to describe various processing techniques such as DMO and slant stacking.

Chapter 6
REVIEW OF MAIN POINTS

A large number of subjects concerning seismic data acquisition and seismic data processing have been discussed here.

The description of multiple-coverage data, gathered for a 2-D seismic line, as a three-dimensional wavefield is of prime importance to a proper understanding of seismic data acquisition and intimately related seismic data processing problems.

This wavefield can be described either in terms of shot coordinate x_s, receiver coordinate x_r and time t or in terms of midpoint coordinate x_m, offset x_o and time t. A simple coordinate transformation links (x_s, x_r) to (x_m, x_o). A number of properties of this wavefield have been described in terms of (t, x_s, x_r) and (t, x_m, x_o).

A minimum spatial sampling interval (called the basic sampling interval) exists that is the same for sampling as a function of x_s and as a function of x_r. The basic sampling interval depends on the maximum recorded frequency and on a minimum apparent velocity V_{min}. V_{min} varies from area to area, hence some areas may need closer sampling than other areas.

Patterns are a bad type of antialias filter. The ultimate in 2-D single-component data acquisition is reached if, for a given desired maximum frequency, the basic sampling interval is used both for shots and receivers without the application of shot and receiver patterns.

Full-resolution recording, defined as shooting and recording at the basic *signal* sampling interval, while using the symmetry criterion for shot and receiver patterns, is the next best technique. Full-resolution recording is an essential method in the case of:

(1) very rough topography or,
(2) wild subsurface (thrusts, karsts) or,
(3) need to reach the ultimate in lateral resolution.

When this method is still too expensive to be justified a careful investigation of the acquisition method may still lead to significant improvements. In marine data acquisition improvement can often be reached by using a proper shot pattern, especially in areas with much diffraction energy, e.g., close to salt domes and in other areas with large lateral velocity contrasts. Shot patterns are equally necessary as receiver patterns. Shot and receiver patterns should be equal in length and ideally, should have the same number of elements. Such symmetric recording leads to optimal results. Adjacent patterns are sufficient, as high-cut k filtering in processing can match the response of overlapping patterns.

Proper sampling in x_s and x_r leads to undersampling in x_m and x_o in the case of off-end data acquisition. A suitable dealiasing method should then be employed. On land this undersampling can be avoided by center-spread shooting with shot stations half-way between receiver stations.

The data processing needs to take the three-dimensional nature of the multiple-coverage data into account by carrying out the various processing steps in a "space-conscious" way. This approach will normally lead to a clearly defined diagnostic phase, completely separated from the application phase by an intermediate determination of parameters for the application.

An important requirement for optimum results, particularly for amplitude measurements on the final stack, is a proper shot- and receiver-strength equalization at the beginning of the processing sequence.

(f,k) filtering is a powerful process. Analogous to shot and receiver patterns, (f,k) filtering must be applied in both (t,x_s) and (t,x_r) or in (t,x_m) and (t,x_o) in order to properly simulate a three-dimensional filter that is actually needed for the three-dimensional wavefield. Only linear x_m independent events can be removed by one (f,k) filter. This filter should be applied in the CMP to avoid smearing in x_m. The (f,k) filter in (t,x_o) should be applied after proper dealiasing. The (f,k) filter in (t,x_m) can be longer than in the other panels and hence more powerful.

The determination of statics may be improved by using comparisons with the nearest neighbors of a trace in all four space-time panels.

Common time panels are a powerful tool to analyze variations of the wavefield in the spatial coordinates.

Stacking after NMO correction attempts to create a zero-offset section of the primaries of one wave type. This technique is not ideal but can be improved by DMO correction.

The equivalence theorem once again illustrates the importance of the three-dimensional description of the wavefield of 2-D multiple-coverage seismic data and also shows the effect of the stacking process on operations performed before stacking.

Compromises in the field technique should be made only after considerations of the requirements of full-resolution recording and space-conscious processing and after consideration of the geological problem to be solved by the geophysical techniques.

Eventually, the rigorous principles that are generally applied in time-domain processes, are to be applied to spatial sampling and spatial processes as well in order to achieve the best possible product with the seismic method.

References

Allen, S. J., 1988, First ten years of 1,024-channel seismic recording: A review: 58th Ann. Internat. Mtg., Soc. Expl. Geophys., Expanded Abstracts, 86–87.

Anstey, N., 1986a, Whatever happened to groundroll?: Leading Edge, **5,** no. 3, 40–45.

Anstey, N., 1986b, Field techniques for high resolution: Leading Edge, **5,** no. 4, 26–34.

Anstey, N., 1986c, A reply by Nigel Anstey: Leading Edge, **5,** no. 12, 19–21.

Anstey, N., 1987, Reply: Leading Edge, **6,** 32 and 48.

Anstey, N., 1989, Stack-array discussion continues: Leading Edge, **8,** 24, 25, and 31.

Backus, M., 1987, Amplitude versus offset: A review: 57th Ann. Internat. Mtg. Soc. Expl. Geophys., Expanded Abstracts, 359–364.

Bardan, V., 1987, Trace interpolation in seismic data processing: Geophys. Prosp. **35,** 343–358.

de Bazelaire, 1988, Normal moveout revisited: Inhomogeneous media and curved interfaces: Geophysics, **53,** 143–157.

Berkhout, A. J., 1984, Seismic resolution, resolving power of acoustical echo techniques: Handbook of geophysical exploration, Seismic exploration, Volume 12, Geophysical Press.

Berkhout, A. J., 1986, The seismic method in the search for oil and gas: Current techniques and future developments: Proc. Inst. Electr. Electron. Eng. **74,** 1133–1159.

Berni, A. J., and Roever, W. L., 1989, Field array performance: Theoretical study of spatially correlated variations in amplitude coupling and static shift and case study in the Paris Basin: Geophysics, **54,** 451–459.

Berryhill, J. R., 1979, Wave equation datuming: Geophysics, **44,** 1329–1339.

Berryhill, J. R., 1986, Submarine canyons: Velocity replacement by wave-equation datuming before stack: Geophysics, **51,** 1572–1579.

Bojarski, N. N., 1983, Generalized reaction principles and reciprocity theorems for the wave equations, and the relationship between the time-advanced and time-retarded fields: J. Acoust. Soc. Am., **74,** 281–285.

Bolondi, G., Loinger, E., and Rocca, F., 1982, Offset continuation of seismic sections: Geophys. Prosp. **30,** 813–828.

Brink, M., and Svendsen, M., 1987, Marine seismic exploration using vertical receiver arrays: acquisition in bad weather: Presented at the 49th Mtg. Euro. Assoc. Expl. Geophys.

Brummit, J. G., 1989, Source directivity and its effect on resolution and signature deconvolution: First Break, **7,** 17–24.

Cassano, E., and Rocca, F., 1974, After-stack multichannel filters without mixing effects: Geophys. Prosp., **22,** 330–344.

Claerbout, J. F., 1985, Imaging the Earth's Interior: Blackwell Scientific Publ.

Clement, W. G., 1973, Basic principles of two-dimensional digital filtering: Geophys. Prosp., **22,** 330–344.

Deregowski, S. M., and Rocca, F., 1981, Geometrical optics and wave theory of constant offset sections in layered media: Geophys. Prosp. **29**, 384–406.

Diebold, J. B., and Stoffa, P. L., 1981, The traveltime equation, tau-p mapping, and inversion of common midpoint data: Geophysics, **46**, 238–254.

Dunkin, J. W., and Levin, F. K., 1973, Effect of normal moveout on a seismic pulse: Geophysics, **38**, 635–642.

Embree, P., Burg, J. P., and Backus, M. M., 1963, Wide-band velocity filtering—The pie-slice filter: Geophysics, **28**, 948–974.

Engstrøm, B., and Hvidsten, J., 1986, Equivalence principles for pre- and poststack multichannel filtering: Presented at the 48th Mtg. Euro. Assoc. Expl. Geophys.

Fail, J. P., and Grau, G., 1963, Les filtres en éventail: Geophys. Prosp., **11**, 131–163.

Fatti, J. L., 1987, No title: Leading Edge, **6**, No. 7, 30–32.

Hale, D., 1984, Dip moveout by Fourier transformation: Geophysics, **49**, 741–757.

Hampson, G., 1987, The relationship of pre-stack apparent velocity filtering to the symmetry of the CMP stack response: First Break, **5**, 359–377.

Hatton, L., Worthington, M. H., and Makin, J., 1987, Seismic data processing. Theory and practice: Blackwell Scientific Publ.

Hobson, M. R., 1985, Shoot between the group for simulated multichannel processing flexibility: First Break, **3**, No. 10, 9–16.

Holzman, M., 1963, Chebyshev optimized geophone arrays: Geophysics, **23**, 145–155.

Hoskin, J. W. J., 1988, Ricker wavelets in their various guises: First Break, **6**, 24–33.

Jerri, A. J., 1977, The Shannon sampling theorem—Its various extensions and applications: A tutorial review: Proc., Inst. Electr. Electron. Eng., **65**, 1565–1596.

Judson, D. R., Schultz, P. S., and Sherwood, J. W. C., 1978, Equalizing the stacking velocities of dipping events via Devilish: Presented at the 48th Ann. Internat. Mtg., Soc. Expl. Geophys.

Kelamis, P. G., and Mitchell, A. R., 1989, Slant-stack processing: First Break, **7**, 43–54.

Knapp, R. W., 1985, Using half-integer source offset with split spread CDP seismic data: Leading Edge, **4**, No. 10, 66–69, 108.

Knopoff, L., and Gangi, A. F., 1959, Seismic reciprocity: Geophysics, **24**, 681–691.

Krohn, C. E., 1984, Geophone ground coupling: Geophysics, **49**, 722–731.

Larner, K., Chambers, R., Yang, M., Lynn, W., and Wai, W., 1983, Coherent noise in marine seismic data: Geophysics, **48**, 854–886.

Lofthouse, J. H., and Bennett, G. T., 1978, Extended arrays for marine seismic acquisition: Geophysics, **43**, 3–22.

Lynn, W., and Larner, K., 1989, Effectiveness of wide marine seismic source arrays: Geophys. Prosp., **37**, 181–207.

Marcoux, M. O., 1981, On the resolution of statics, structure, and residual moveout: Geophysics, **46**, 984–993.

Marschall, R., 1980, Zweidimensionale Rekursivfilter, in: Festschrift Theodor Krey, Prakla Seismos GmbH.

Mayne, W. H., 1962, Common reflection point horizontal stacking techniques: Geophysics, **27**, 927–938.

Mitchell, A. R., and Kelamis, P. G., 1990, Efficient tau-p hyperbolic filtering: Geophysics, **55**, 619–625.

Morgan, N. A., 1970, Wavelet maps: A new analysis tool for reflection seismograms: Geophysics, **35**, 447–460.

Morse, P. F., and Hildebrandt, G. F., 1989, Ground-roll suppression by the stack-array: Geophysics, **54**, 290–301.

Newman, P., 1984, Seismic response to sea-floor diffractors: First Break, **2**, (2), 9–19.

Newman, P., and Mahoney, J. T., 1973, Patterns—with a pinch of salt: Geophys. Prosp. **21**, 197–219.

Nyquist, H., 1928, Certain topics in telegraph transmission theory: Trans., Am. Inst. Elec. Eng., **47**, 617–644.

Ongkiehong, L., and Askin, H., 1988, Towards the universal seismic acquisition technique: First Break, **6**, 46–63.

Ongkiehong, L., and Huizer, W., 1987, Dynamic range of the seismic system: First Break, **5**, 435–439.

Oppenheim, A. V., and Schafer, R. W., 1975, Digital signal processing: Prentice-Hall, Inc.

Ottolini, R., and Claerbout, J. F., 1984, The migration of common midpoint slant stacks: Geophysics, **49**, 237–249.

Petersen, D. P., and Middleton, D., 1962, Sampling and reconstruction of wave-number limited functions in N-dimensional Euclidean spaces: Information and Control, **5**, 279–323.

Rietsch, E., 1980, Estimation of the S/N ratio of seismic data with an application to stacking: Geophys. Prosp. **28**, 531–550.

Robinson, J. C., 1970, Statistically optimal stacking of seismic data: Geophysics, **35**, 436–446.

Rocksandić, M. M., 1986, Extended marine arrays versus simulated extended arrays: Geoph. Prosp, **34**, 1154–1166.

Ronen, J., 1987, Wave-equation trace interpolation: Geophysics, **52**, 973–984.

Schultz, P. S., 1982, A method for direct estimation of interval velocities: Geophysics, **47**, 1657–1671.

Schultz, P. S., Pieprzak, A. W., and Loh, E. K. L., 1983, A case for larger offsets: Geophysics, **48**, 238–247.

Shannon, C. E., 1949, Communication in the presence of noise: Proc., Inst. Radio Eng., **37**, 10–21.

Sheriff, R. E., 1984, Encyclopedic dictionary of exploration geophysics: Soc. Expl. Geophys.

Sheriff, R. E., and Geldart, L. P., 1982, Exploration seismology, Volume 1, History, theory, and data acquisition: Cambridge Univ. Press.

Sheriff, R. E., and Geldart, L. P., 1983, Exploration seismology, Volume 2: Data-processing and interpretation: Cambridge Univ. Press.

Stoffa, P. L., Buhl, P., Diebold, J. B., and Wenzel, F., 1981, Direct mapping of seismic data to the domain of intercept time and ray parameters—A plane-wave decomposition: Geophysics, **46**, 255–267.

Taner, M. T., 1976, Simplan-simulated plane-wave exploration: Presented at the 46th Ann. Internat. Mtg., Soc. Expl. Geophys.

Taner, M. T., Koehler, F., and Alhilali, K. A., 1974, Estimation and correction of near-surface time anomalies: Geophysics, **39**, 441–463.

Taner, M. T., and Koehler, F., 1981, Surface consistent corrections: Geophysics, **46**, 17–22.

Ursin, B., 1978, Attenuation of coherent noise in marine seismic exploration using very long arrays: Geophys. Prosp., **26**, 722–749.

Ursin, B., 1983, Spatial filtering of marine seismic data: Geophysics, **48**, 1611–1630.

White, J. E., 1960, Use of reciprocity theorem for computation of low frequency radiation patterns: Geophysics, **25**, 613–624.

White, R. E., 1977, The performance of optimum stacking filters in suppressing uncorrelated noise: Geophys., Prosp., **25**, 165–178.

Yilmaz, O., and Claerbout, J. F., 1980, Prestack partial migration: Geophysics, **45**, 1753–1779.

Yilmaz, O., 1987, Seismic data processing: Investigations in geophysics No. 2, Soc. Expl. Geophys.

Index